Lecture Notes in Geoinformation and Cartography

Series Editors: William Cartwright, Georg Gartner, Liqiu Meng, Michael P. Peterson

For further volumes:
http://www.springer.com/series/7418

Wenzhong Shi • Anthony G.O. Yeh •
Yee Leung • Chenghu Zhou
Editors

Advances in Spatial Data Handling and GIS

14th International Symposium on Spatial Data Handling

Springer

Editors
Wenzhong Shi
Department of Land Surveying and Geo-Informatics
The Hong Kong Polytechnic University
Hong Kong SAR
China

Anthony G.O. Yeh
Centre of Urban Studies and Urban Planning
Department of Urban Planning and Design
The University of Hong Kong
Hong Kong SAR
China

Yee Leung
Department of Geography and Resource Management
The Chinese University of Hong Kong
Hong Kong SAR
China

Chenghu Zhou
Chinese Academy of Sciences
Institute of Geographical Sciences and Natural Resources Research
Beijing
China

ISSN 1863-2246 ISSN 1863-2351 (electronic)
ISBN 978-3-642-25925-8 ISBN 978-3-642-25926-5 (eBook)
DOI 10.1007/978-3-642-25926-5
Springer Heidelberg New York Dordrecht London

Library of Congress Control Number: 2012937841

Printed on acid-free paper

Springer is part of Springer Science+Business Media (www.springer.com)

Foreword

This book is the selected papers of the *14th International Symposium on Spatial Data Handling* that was held at The Hong Kong Polytechnic University in Hong Kong from 26 to 28 May 2010 as a *Joint International Conference on Theory, Data Handling and Modelling in GeoSpatial Information Science* organized by Commission on Geographical Information Science and Commission on Modelling Geographical Systems of the International Geographical Union (IGU) and Commission II of the International Society of Photogrammetry and Remote Sensing (ISPRS).

The *International Symposium on Spatial Data Handling (SDH)* is the premier biennial international research forum for Geographical Information Science (GISc) of the Commission on Geographical Information Science of IGU. It was first held in 1984 in Zurich, Switzerland, when the Commission was named as Geographical Data Sensing and Processing chaired by Prof. Duane Marble. It has quickly became the premier scientific meetings in the GIS area and since then been held in Seattle, USA; Sydney, Australia; Zurich, Switzerland; Charleston, USA; Edinburgh, UK; Delft, The Netherlands; Vancouver, Canada; Beijing, China; Ottawa, Canada; Leicester, UK; Vienna, Austria; and Montpellier, France.

The IGU Commission on Modeling Geographical Systems has been organizing its conferences over the years on the modeling and analysis of geographical data and systems in different parts of the world. It is the first time that it joins hand with ISPRS and IGU Commission on Geographical Information Science to co-organize the conference to encourage interaction and collaboration among researchers and professionals in the three organizations.

ISPRS is a society regrouping scientific societies from more than 100 countries working in domains related to photogrammetry, remote sensing and geographical information science. The ISPRS Technical Commission II Symposium is organised every 4 years alternately with the ISPRS Congress and is among the major events in ISPRS calendar regrouping leading scholars from the GISc and related communities. Past symposiums were held in Ottawa, Canada (2002) and Vienna, Austria (2006) jointly with SDH.

The main objective is to bring together scholars and professionals from the international community of GISc to present the latest achievements and to share experience in GISc research. The conference programme featured keynote speeches delivered by leading GISc scholars, technical sessions with reports of the latest research outcomes, student forum on meeting with GISc legends and commercial exhibitions showing the latest development of GISc technology.

Acknowledgements

We would like to thank the financial support of the Faculty of Construction and Land Use, The Hong Kong Polytechnic University; Chung Chi College and the Faculty of Social Science, The Chinese University of Hong Kong; Department of Geography, Hong Kong Baptist University; and the State Key Laboratory of Resources and Environment Information System, Institute of Geographic, Sciences and Natural Resources Research, Chinese Academy of Sciences in organizing the *Joint International Conference on Theory, Data Handling and Modelling in GeoSpatial Information Science* that was held at The Hong Kong Polytechnic University, Hong Kong, on 26–28 May 2010.

We would like to thank Mr. Zelang Miao from China University of Mining and Technology in the contribution in producing this book.

Contents

Programme Committee

Chair

Yee Leung, *The Chinese University of Hong Kong, Hong Kong SAR, China*

Co-Chairs

Brian Lees, *University of New South Wales, Australia*
Wenzhong Shi, *The Hong Kong Polytechnic University, Hong Kong SAR, China*
Qiming Zhou, *Baptist University of Hong Kong, Hong Kong SAR, China*

Members

Masatoshi Arikawa, *The University of Tokyo, Japan*
Özgün Akçay, *Istanbul Technical University, Turkey*
Ali Bennasr, *University of Sfax, Tunisia*
Jean-Paul Bord, *University of Montpellier III Paul Valéry, France*
Tao Cheng, *University College London, UK*
Christophe Claramunt, *Naval Academy Research Institute, France*
Graham Clarke, *University of Leeds, UK*
Hande Demirel, *Istanbul Technical University, Turkey*
Manfred M. Fischer, *Vienna University of Economics and Business, Austria*
Stewart Fotheringham, *National University of Ireland, Maynooth, Ireland*
Andrew Frank, *Vienna University of Technology, Austria*
Tung Fung, *The Chinese University of Hong Kong, Hong Kong SAR, China*
Christopher Gold, *University of Glamorgan, UK*
Michael Frank Goodchild, *University of California, Santa Barbara, USA*

Daniel A. Griffith, *The University of Texas at Dallas, USA*
Lars Harrie, *Lund University, Sweden*
Francis Harvey, *University of Minnesota, USA*
Andy Hudson-Smith, *University College London, UK*
Gerhard Joos, *Technical University of Denmark, Denmark*
Pavlos Kanaroglou, *McMaster University, Canada*
Marinos Kavouras, *National Technical University of Athens, Greece*
Zhilin Li, *The Hong Kong Polytechnic University, Hong Kong SAR, China*
Hui Lin, *The Chinese University of Hong Kong, Hong Kong SAR, China*
Yaolin Liu, *Wuhan University, China*
Shattri Mansor, *Putra University, Malaysia, Malaysia*
Lingkui Meng, *Wuhan University, China*
Martien Molenaar, *University of Twenthe, The Netherlands*
Mir Abolfazl Mostafavi, *Laval University, Canada*
Christopher J. Pettit, *Department of Primary Industries, Australia*
Alias Abdul Rahman, *Technological University of Malaysia, Malaysia*
Docent Juri Roosaare, *University of Tartu, Estonia*
Anne Ruas, *University of Marne-la-Vallée, France*
Yukio Sadahiro, *The University of Tokyo, Japan*
Mauro Salvemini, *Sapienza University of Rome, Italy*
Monika Sester, *University of Hanover, Germany*
A. Rashid B. M. Shariff, *Putra University, Malaysia, Malaysia*
Ali Sharifi, *University of Twenthe, The Netherlands*
Ryosuke Shibasaki, *The University of Tokyo, Japan*
Nadine Schuurman, *Simon Fraser University, Canada*
Ake Sivertun, *Linköping University, Sweden*
Terese Steenberghen, *University of Leuven, Belgium*
Alfred Stein, *University of Twente, The Netherlands*
Xinming Tang, *Chinese Academy of Surveying and Mapping, China*
Vladimir Tikunov, *Moscow State University, Russia*
Abdülvahit Torun, *University of Minnesota, Turkey*
Vit Vozenilek, *Palacký University of Olomouc, Czech Republic*
Jinfeng Wang, *Chinese Academy of Sciences, China*
Shuliang Wang, *Wuhan University, China*
Robert Weibel, *University of Zurich, Switzerland*
Stephan Winter, *University of Melbourne, Australia*
Michael Worboys, *The University of Maine, USA*
Anthony G.O. Yeh, *The University of Hong Kong, Hong Kong SAR, China*
Chengu Zhou, *Institute of Geographical Sciences and Natural Resources Research, China*

Local Organising Committee

Chair

Wenzhong Shi, *The Hong Kong Polytechnic University, Hong Kong SAR, China*

Co-Chairs

Yee Leung, *The Chinese University of Hong Kong, Hong Kong SAR, China*
Chenghu Zhou , *Chinese Academy of Sciences, China*
Qiming Zhou, *Baptist University of Hong Kong, Hong Kong SAR, China*

Secretary

Eric Guilbert, *The Hong Kong Polytechnic University, Hong Kong SAR, China*

Members

Bo Huang, *The Chinese University of Hong Kong, Hong Kong SAR, China*
Brian Lees, *University of New South Wales, Australia*
Cheuk Man Lung, *Baptist University of Hong Kong, Hong Kong SAR, China*
Bo Wu, *The Hong Kong Polytechnic University, Hong Kong SAR, China*
Anthony G.O. Yeh, *The University of Hong Kong, Hong Kong SAR, China*

Commission on Geographical Information Science

International Geographical Union (IGU)

Chair

Brian Lees, *University of New South Wales, Australia*

Vice-Chair

Chengu Zhou, *Institute of Geographical Sciences and Natural Resources Research, Chinese Academy of Sciences, China*

Members

Itzhak Benenson, *Tel-Aviv University, Israel*
Jean-Paul Bord, *University of Montpellier III Paul Valéry, France*
Francis Harvey, *University of Minnesota, USA*
Thangavelu Kumaran, *University of Madras, India*
Henk Ottens, *Utrecht University, Netherlands*
Docent Juri Roosaare, *University of Tartu, Estonia*
Mauro Salvemini, *University of Rome "La Sapienza", Italy*
Nadine Schuurman, *Simon Fraser University, Canada*
Abdul Rashid Bin Mohamed Shariff, *Universiti Putra Malaysia, Malaysia*
Ake Sivertun, *Linkoping University, Sweden*
Vladimir Tikunov, *Moscow State University, Russia*
Vit Vozenilek, *Palacky University, Czech Republic*

Commission on Modelling Geographical Systems

International Geographical Union (IGU)

Chair

Yee Leung, *The Chinese University of Hong Kong, Hong Kong, SAR, China*

Vice-Chair

Stewart Fotheringham, *National University of Ireland, Ireland*

Secretary

Pavlos Kanaroglou, *McMaster University, Canada*

Members

Ali Bennasr, *University of Sfax, Tunisia*
Graham Clarke, *University of Leeds, UK*
Manfred M Fischer, *Vienna University of Economics and Business, Austria*
Daniel A. Griffith, *University of Texas at Dallas, USA*
Yukio Sadahiro, *University of Tokyo, Japan*
Therese Steenberghen, *University of Leuven, Belgium*
Jinfeng Wang, *Institute of Geography, Chinese Academy of Sciences, China*

Commission II: Theory and Concepts of Spatial Information Science

International Society of Photogrammetry and Remote Sensing (ISPRS)

President

Wenzhong Shi, *The Hong Kong Polytechnic University, Hong Kong SAR, China*

Secretary

Eric Guilbert, *The Hong Kong Polytechnic University, Hong Kong SAR, China*

WG II/1 - Cognition and modeling of space and time

Chair

Xinming Tang, *Chinese Academy of Surveying and Mapping, China*

Co-Chair

Stephan Winter, *The University of Melbourne, Australia*
Lingkui Meng, *The Wuhan University, China*

Secretary

Bisheng Yang, *Wuhan University, China*

WG II/2 - Multi-scale representation of spatial data

Chair

Monica Sester, *Leibniz Universität Hannover, Germany*

Co-Chair

Lars Harrie, *Lund University, Sweden*

Secretary

Jan Haunert, *Leibniz Universität Hannover, Germany*

WG II/3 - Spatial analysis and data mining

Chair

Yaolin Liu, *Wuhan University, China*

Co-Chair

Tao Cheng, *University College London, United Kingdom*
Abdulvahit Torun, *University of Minnesota, United States*

Secretary

Shuliang Wang, *Wuhan University, China*

WG II/4 - Uncertainty modeling and quality control for spatial data

Chair

Alfred Stein, *ITC, The Netherlands*

Co-Chair

Gerhard Joos, *Technical University of Denmark, Denmark*
Hande Demirel, *Istanbul Technical University, Turkey*

Secretary

M.R. Delavar, *University of Tehran, Iran*

WG II/5 - Multidimensional and mobile data models

Chair

Alias Abdul Rahman, *University of Technology Malaysia, Malaysia*

Co-Chair

Christopher Gold, *University of Glamorgan, United Kingdom*
François Anton, *Technical University of Denmark, Denmark*

Secretary

Pawel Boguslawski, *Universiti Teknologi Malaysia, Malaysia*

WG II/6 - Geo-visualization and Virtual Reality

Chair

Christopher J. Pettit, *The University of Melbourne, Australia*

Co-Chair

Andy Hudson-Smith, *University College London, United Kingdom*
Bo Wu, *The Hong Kong Polytechnic University, Hong Kong*

Secretary

Arzu Çöltekin, *University of Zurich, Switzerland*

WG II/7 - Spatial decision support and location-based services

Chair

Ali Sharifi, *ITC, The Netherlands*

Co-Chair

Shattri Mansor, *University of Putra Malaysia, Malaysia*
Masatoshi Arikawa, *University of Tokyo, Japan*

Secretary

Rainer Laudien, *University of Cologne, Germany*

WG II/IV - Semantic data interoperability and ontology for geospatial information

Chair

Ryosuke Shibasaki, *University of Tokyo, Japan*

Co-Chair

Marinos Kavouras, *National Technical University of Athens, Greece*
Mir Abolfazl Mostafavi, *Laval University, Canada*

Secretary

Margarita Kokla, *National Technical University of Athens, Greece*

Part I
Modeling Space and Time

Chapter 1
Deriving Space-Time Variograms from Space-Time Autoregressive (STAR) Model Specifications

Daniel A. Griffith and Gerard B.M. Heuvelink

Abstract Many geospatial science subdisciplines analyze variables that vary over both space and time. The space–time autoregressive (STAR) model is one specification formulated to describe such data. This paper summarizes STAR specifications that parallel geostatistical model specifications commonly used to describe space–time variation, with the goal of establishing synergies between these two modeling approaches. Resulting expressions for space–time correlograms derived from 1st-order STAR models are solved numerically, and then linked to appropriate space–time semivariogram models.

1.1 Introduction

Geostatistics furnishes techniques for modeling the covariance matrix, whereas spatial autoregression furnishes techniques for modeling the inverse covariance matrix, for a set of n geographically distributed values of a single random function. Both seek to capture spatial autocorrelation effects in georeferenced data.

Although, in practice, both geostatistics and spatial autocorrelation techniques mostly are applied to static spatial variables, a growing interest among researchers is to utilize these techniques to address change over both space and time. Incorporating time is more than just adding another dimension, because the behavior of a variable over time differs from its behavior over space, and characteristics of temporal processes often are known to some degree. Accordingly, a space–time geostatistical or autoregressive model must capture the fundamental differences

D.A. Griffith (✉)
School of EPPS, University of Texas at Dallas, 800 W. Campbell Road, GR 31, 75080 Richardson, TX, USA
e-mail: dagriffith@utdallas.edu

G.B.M. Heuvelink
Wageningen University and Research Centre, Wageningen, The Netherlands

A. Yeh et al. (eds.), *Advances in Spatial Data Handling and GIS*,
Lecture Notes in Geoinformation and Cartography,
DOI 10.1007/978-3-642-25926-5_1,

between spatial and temporal variation, and must include these differences in its structure and parameterization.

The purpose of this paper is to establish the basis for a synergy between space–time geostatistics and autoregressive (STAR) approaches to the modeling of correlation structures latent in space–time data. The mutually advantageous conjunction of these two approaches follows that established for the static case by Griffith and Csillag (1993) and Griffith and Layne (1997), and seeks to create an enhanced combined approach to the modeling of space–time correlation structures. Simple 1st- and 2nd-order geographic neighbor direct dependency structures are addressed, with conceptualizations furnished by especially Gasim (1988) allowing them to be extended to larger neighborhoods. In doing so, we exploit the notion that a space–time semivariogram is valid (i.e., non-negative definite) when any linear combination of values of the associated random function at any finite number of space–time points has non-negative variance.

1.2 The Configurational Structure of Georeferenced Data

Consider a variable $Z = \{Z(s,t)|s \in S, t \in T\}$ that varies within a spatial domain S and a time horizon T. Let Z be observed at n space–time points (s_i, t_j), $i = 1, 2, \ldots, m$ and $j = 1, 2, \ldots, k$, where $n = mk$. These space–time observations constitute a time series of length k at each of the m spatial locations, and imply the use of a regular sampling scheme (i.e., the observations are uniformly spaced over time at each spatial location).

The set of m points can be converted to a surface partitioning by constructing its associated set of Thiessen (Voronoi) polygons; these become volumes in three dimensions. This conversion allows the generation of a Delauany triangulation (the dual graph) that furnishes a topology-based articulation of the configurational structure of the set of m points. Interpoint distances furnish another. Suppose variable Z is an areal unit aggregate observed for m regions in time, where k is the frequency of observations per region over time. Let these areal units form a mutually exclusive and collectively exhaustive partitioning of a surface. If these polygons are convex hulls (all internal angles $<180°$), then the geometric centroid of each polygon can be computed, and this set of points can be used both to convert the surface partitioning into a geographic distribution of points, and to construct a dual graph (similar to a Delaunay triangulation). This graph commonly is constructed using criteria based on chess piece movements: the rook's case (i.e., links connect points for polygons that share a common non-zero length boundary), and the queen's case (i.e., links connect points for polygons that share a common zero—i.e., point—or non-zero length boundary). For concave hulls (e.g., polygons with at least one internal angle $>180°$) or for nested areal units (e.g., one contained completely inside another), a judiciously selected arbitrary point may be the dual graph node. Meanwhile, time can be represented with a simple line graph comprising a linear sequence of links and points.

In all three geographic cases, the graphs in question can be converted to adjacency matrices, C, which are binary 0–1 matrices with all diagonal entries being 0. Because these graphs are planar or near-planar and connected, the number of ones in the m-by-m matrices representing geographic arrangement is at least 2(m-1), usually does not exceed 3(m-2), and never exceeds 8m. These matrices are symmetric here, in part because geographic dependencies are being cast as non-directional. The number of ones in the k-by-k time sequencing matrix is 2(k-1). This set of matrices furnishes the building blocks of n-by-n space–time data matrices. Eigenfunctions extracted from each of these binary matrices can be used to summarize their respective structure.

1.3 STAR Model Specifications

STAR model specifications (see Cliff et al. 1975) are explicit formulations describing how a variable Z varies in space s = (x, y) and time (t) in some joint fashion (x, y, t). The following two linear discrete cases are of interest here:

$$\begin{aligned} Z(x,y,t) = {} & a \times Z(x,y,t-\Delta t) + b \times \{Z(x-\Delta x,y,t-\Delta t) + Z(x+\Delta x,y,t-\Delta t) \\ & + Z(x,y-\Delta y,t-\Delta t) + Z(x,y+\Delta y,t-\Delta t)\} + \varepsilon(x,y,t)\,, \end{aligned} \tag{1.1}$$

and

$$\begin{aligned} Z(x,y,t) = {} & a \times Z(x,y,t-\Delta t) + b \times \{Z(x-\Delta x,y,t) + Z(x+\Delta x,y,t) \\ & + Z(x,y-\Delta y,t) + Z(x,y+\Delta y,t)\} + \varepsilon(x,y,t). \end{aligned} \tag{1.2}$$

Equation (1.1) specifies a value at location (x, y, t) as a function of the preceding in situ value (time t-Δt) as well as the preceding neighboring values, a lagged specification. Equation (1.2) specifies a value at location (x, y, t) as a function of the preceding in situ value (time t-Δt) as well as the contemporaneous neighboring values, a spatially contemporaneous specification. The random process ε is Gaussian white noise, which is uncorrelated in space and time. In the STAR model, correlation in space and time is captured by the autoregressive structure of the model (i.e., the response variable appears in both sides of the equations). Feedback loops or cycles make (1.2) fundamentally different from (1.1). An initial field for t = 0 and spatial boundary conditions are needed in these formulations. In this paper, interest is in the case where sufficient time has transpired and the spatial extent is sufficiently large to allow negligible effects from boundary conditions.

1.3.1 Theoretical Space–Time Correlations

Theoretical correlations can be posited for (1.1) and (1.2). Consider an infinite regular square (i.e., equal-sized pixels) tessellation lattice for which spatial adjacency (i.e., geographic neighbors) is defined by whether or not two square cells share a non-zero length common boundary (i.e., the rook's definition). Let {Z(s)} be a Gaussian random variable distributed across the vector of locations s (i.e., cells), such that {Z(s)} and {Z(s + h)}, for locations shifted by h units, are stochastically equivalent (i.e., complete stationarity). Spectral theory (Bartlett 1975; Haining 1978) reveals that the appropriate correlation function for the additive specification [i.e., (1.1)] is given by

$$\rho_{h,g,k} = \frac{\int_0^\pi\int_0^\pi\int_0^\pi \frac{\cos(h\times t)\cos(g\times u)\cos(k\times v)}{\{1-\rho_T\cos(t)-\rho_s\cos(u)-\rho_s\cos(v)\}^\eta}\,dt\,du\,dv}{\int_0^\pi\int_0^\pi\int_0^\pi 1/\{1-\rho_T\cos(t)-\rho_s\cos(u)-\rho_s\cos(v)\}^\eta dt\,du\,dv}, \tag{1.3}$$

whereas that for the multiplicative specification [i.e., (1.2)] is given by

$$\rho_{h,g,k} = \frac{\int_0^\pi\int_0^\pi\int_0^\pi \frac{\cos(h\times t)\cos(g\times u)\cos(k\times v)}{[1-\cos(t)\{\rho_s[\cos(u)+\cos(v)]+\rho_T\}]^\eta}\,dt\,du\,dv}{\int_0^\pi\int_0^\pi\int_0^\pi 1/[1-\cos(t)\{\rho_s[\cos(u)+\cos(v)]+\rho_T\}]^\eta dt\,du\,dv}, \tag{1.4}$$

for temporal lag h (h = 0, 1, ...), and spatial lags g (g = 0, 1, ...) and f (f = 0, 1, ...), where a positive integer value of η yields an η th-order model, ρ_s is the spatial and ρ_T is the temporal autoregressive parameter, and (h, g, f) denotes the space–time lag involved.

1.3.2 Space–Time Autoregressive Structures

The eigenvalues of the n-by-n connectivity matrix **C** for a linear surface partitioning containing P cells are $2\cos(\frac{p\pi}{P+1})$, p = 1, 2, ..., P. The 2 can be absorbed into the autoregressive parameter values, ρ_j, doubling the size of each feasible parameter space. This solution can be extended to two- and three-dimensional regular square lattice structures. Ord (1975) first reported the eigenvalues of the PQ-by-PQ connectivity matrix for a square tessellation surface partitioning forming a P-by-Q (n = PQ) complete rectangular region as $2[\cos(\frac{p\pi}{P+1}) + \cos(\frac{q\pi}{Q+1})]$, p = 1, 2, ..., P, and q = 1, 2, ..., Q. Gasim (1988) presents extensions to Ord's results. And, Basilevsky (1983) summarizes the conventional time-series results. Here the three-dimensional matrix representation is given by

$$\mathbf{C} = \mathbf{I}_T \otimes \mathbf{I}_s - \rho_s \mathbf{I}_T \otimes \mathbf{C}_s - \rho_T \mathbf{C}_T \otimes \mathbf{I}_s,$$

where $\otimes$ denotes Kronecker product, $\mathbf{I}_T$ denotes the k-by-k identify matrix, $\mathbf{I}_s$ denotes the PQ-by-PQ identity matrix, and $\mathbf{C}_s = \mathbf{C}_P \otimes \mathbf{I}_Q + \mathbf{C}_Q \otimes \mathbf{I}_P$, for a P-by-Q rectangular square lattice, where $\mathbf{C}_j$ is a matrix of 0s except for the upper- and lower-off diagonals, which contain 1s (j = P, Q, and T). $\mathbf{C}_P$ and $\mathbf{C}_Q$ have the same structure as $\mathbf{C}_T$.

Equation (1.3) describes the correlogram values for space–time data characterized by (1.1), whereas (1.4) describes space–time data characterized by (1.2). The three-dimensional connectivity matrix representation is given by

$$\mathbf{C} = \mathbf{I}_T \otimes \mathbf{I}_s - \rho_s \mathbf{C}_T \otimes \mathbf{C}_s - \rho_T \mathbf{C}_T \otimes \mathbf{I}_s,$$

where $1 - \cos(t)\{\rho_s[\cos(u) + \cos(v)] + \rho_T\}$ are the limiting eigenvalues of the space–time connectivity matrix **C**. Additional discussion of this topic appears in Griffith (1996).

Because the eigenvalues define the spectrum of a matrix, they appear in the denominator of (1.3) and (1.4); these denominators are based upon the limiting eigenvalues of the connectivity matrix representation of the space–time three-dimensional structure of data. In addition, Griffith and Csillag (1993) note, in contrast to the current thinking of that time, that a simultaneous autoregressive model can be portrayed by letting $\eta = 2$ in the denominator of (1.3) and (1.4)—it becomes a 2nd-order covariance specification; Bartlett (1975, pp. 19, 25) reports this result. Furthermore, Griffith and Layne (1997) summarize the close numerical connections between the spatial-only form of (1.3) and (1.4) and geostatistical semivariogram models.

1.3.3 Space–Time Covariance Functions in Geostatistics

An important issue in the space–time geostatistical literature concerns whether or not the space and the time components of a formulated function are: separable such that they factor (Gneiting et al. 2006); or, nonseparable such that they form a linear combination (Ma 2008). Mitchell et al. (2005) propose a modified multivariate repeated measures likelihood ratio test coupled with bootstrapping for this purpose. Brown et al. (2001) note that separability requires that the expected value for some random variable at location (x, y) in time t + 1, given its values in a neighborhood of location (x, y) in time t, must equal the conditional expectation just for location (x, y).

Stein (2005) furnishes an overview of space–time covariance and aspects of spatial-temporal interaction, and proposes a new class of space–time covariance models. Ma (2003, 2008) presents methods for constructing spatio-temporal stationary covariance models, and supplements the set presented by Kolovos et al. (2004).

Gneiting et al. (2006) posit theorems for symmetric and separable specifications, the Cressie-Huang and the Gneiting model, and stationarity. Fuentes et al. (2008) propose a nonstationary and nonseparable spectral density specification for which separability is a special case. Finally, Calder (2007) proposes a Bayesian specification that includes priors on initial points in time.

The space–time separability assumption (Bogaert 1996) states that the space–time covariance function C(h, u) can be written as a product of a spatial, $C_S(h)$, and a temporal, $C_T(u)$, covariance function, such that

$$C(h,u) = C_S(h) \times C_T(u) \tag{1.5}$$

One non-separable specification expresses the space–time covariance function as a linear combination of these two components (De Cesare et al. 2001), such that

$$C(h,u) = C_S(h) + C_T(u) + pC_S(h) \times C_T(u), \tag{1.6}$$

which is statistically valid if both $C_S(h)$ and $C_T(u)$ are valid covariance functions and parameter p satisfies certain conditions (De Cesare et al. 2001). This product-sum model appears to perform well in practice (De Iaco et al. 2003; Gething et al. 2007).

Another alternative is the metric model (Dimitrakopoulos and Luo 1994), which reduces the space–time covariance function to

$$C(h,u) = C_{ST}(\sqrt{h^2 + (\alpha \times u)^2}), \tag{1.7}$$

whose essential characteristic is that distance in space is made comparable to distance in time through the scaling parameter α. Equation (1.7) is rather restrictive because it assumes that the variances in time and space are equal. The following more flexible specification results from combining (1.6) with p = 0 and (1.7):

$$C(h,u) = C_S(h) + C_T(u) + C_{ST}(\sqrt{h^2 + (\alpha \times u)^2}). \tag{1.8}$$

The third term in the right-hand side represents a joint space–time interaction effect.

1.4 Numerical Experiments

Only numerical integration solutions to the definite integrals in (1.3) and (1.4) are available here. Because this integration is numerical intensive, and 1st- and 2nd-order results are similar, only 1st-order models are assessed. Because the exponent in the denominators of the integrands is 1, (1.1) and (1.2) refer to a space–time conditional autoregressive (CAR) specification. Numerical results for (1.3) and (1.4) were calculated for time lags h = 0, 1, …, 65 and space lags g and f = 0, 1, …, 50, using the autoregressive parameter pairs $\{(\rho_s, \rho_T):$

Table 1.1 Parameter estimates for the exponential variogram model and contemporaneous spatial dependence

ρ_s	ρ_T	Space			Time			a	RESS	Space–time		
		C0	C1	r	C0	C1	r			C0	C1	r
Equation (1.5)												
0.01	0.97	1.2052	0.0992	0.5212	0.7661	0.0006	4.9569		0.8594			
0.10	0.79	1.2315	0.0514	0.7712	0.7788	0.0007	2.7328		0.9279			
0.20	0.59	1.3987	0.0459	0.8321	0.6917	0.0006	2.4328		0.9451			
0.30	0.39	1.1825	0.1004	0.0000	0.7787	0.0008	1.9168		0.9986			
0.40	0.19	1.2671	0.0309	0.8431	0.7697	0.0007	1.6549		0.9620			
0.49	0.01	1.3535	0.0283	0.8041	0.7231	0.0006	1.3169		0.9721			
Equation (1.6)												
0.01	0.97	0.0000	1.0000	0.4543	0.0096	0.9892	4.4235	−1.0000	0.0054			
0.10	0.79	0.0008	0.9992	0.5960	0.0281	0.9685	1.9877	−1.0000	0.0441			
0.20	0.59	0.0000	1.0000	0.5964	0.0270	0.9691	1.3773	−1.0000	0.0684			
0.30	0.39	0.0105	0.9894	0.5675	0.0106	0.9852	1.0190	−1.0000	0.0739			
0.40	0.19	0.0079	0.9920	0.5076	0.0049	0.9907	0.7222	−1.0000	0.0673			
0.49	0.01	0.0006	0.9992	0.2962	0.0000	0.9952	0.3038	−1.0000	0.1187			
Equation (1.7)												
0.01	0.97							0.0078	0.0029	0.0052	0.9948	0.3869
0.10	0.79							0.0714	0.0340	0.0208	0.9792	0.5214
0.20	0.59							0.1585	0.0560	0.0207	0.9793	0.5414
0.30	0.39							0.2784	0.0601	0.0174	0.9826	0.5381
0.40	0.19							0.4727	0.0526	0.0126	0.9874	0.5145
0.49	0.01							0.8494	0.1137	0.0044	0.9956	0.8494
Equation (1.8)[a]												
0.01	0.97	0	***	***	0	0.0014	***	0.0792	0.0029	0	0.9986	0.3869
0.10	0.79	0	0.0001	32.9987	0	0.0029	1.9357	0.0737	0.0337	0	0.9970	0.5135
0.20	0.59	0	0.0001	16.0388	0	0.0034	2.1593	0.1635	0.0552	0	0.9966	0.5317
0.30	0.39	0	0.0001	13.7028	0	0.0035	2.2861	0.2857	0.0588	0	0.9964	0.5280
0.40	0.19	0	0.0001	10.1898	0	0.0037	2.2348	0.4825	0.0509	0	0.9963	0.5046
0.49	0.01	0	0.0002	7.2160	0	0.0044	2.1385	0.8653	0.1112	0	0.9954	0.3806

Note: *** denotes an estimate at the limit of the numerically calculated space–time data cube; C0 denotes the nugget, C1 the sill, and r the range parameter

[a]The three C0 terms were set equal to 0—the theoretical value—in order to achieve convergence

Table 1.2 Parameter estimates for the exponential variogram model and time-lagged spatial dependence

ρ_s	ρ_T	Space			Time			a	RESS	Space–time		
		C0	C1	r	C0	C1	r			C0	C1	r
Equation (1.5)												
0.01	0.97	1.2950	0.1065	0.5216	0.7130	0.0006	4.9405		0.8597			
0.10	0.79	1.3720	0.0563	0.7786	0.6995	0.0007	3.4037		0.9366			
0.20	0.59	1.3902	0.0440	0.8213	0.6966	0.0007	2.0341		0.9525			
0.30	0.39	1.3016	0.0337	0.9219	0.7481	0.0009	1.5630		0.9607			
0.40	0.19	1.3459	0.0292	0.9110	0.7263	0.0009	1.1788		0.9711			
0.49	0.01	1.2671	0.0210	1.0209	0.7750	0.0015	0.4744		0.9816			
Equation (1.6)												
0.01	0.97	0.0000	1.0000	0.4564	0.0092	0.9897	4.4207	−1.0000	0.0054			
0.10	0.79	0.0003	0.9996	0.6219	0.0240	0.9733	1.9716	−1.0000	0.1266			
0.20	0.59	0.0000	0.9999	0.6627	0.1955	0.9779	1.3631	−1.0000	0.1093			
0.30	0.39	0.0000	0.9999	0.6925	0.0122	0.9856	1.0026	−1.0000	0.0650			
0.40	0.19	0.0039	0.9961	0.7478	0.0000	0.9986	0.7322	−1.0000	0.0874			
0.49	0.01	0.0000	0.9663	0.7875	0.8039	0.1919	0.2914	−0.9999	0.7325			
Equation (1.7)												
0.01	0.97							0.0079	0.0030	0.0047	0.9953	0.3880
0.10	0.79							0.0788	0.1190	0.0147	0.9853	0.5386
0.20	0.59							0.2001	0.1008	0.0104	0.9896	0.5885
0.30	0.39							0.4346	0.0575	0.0044	0.9955	0.6321
0.40	0.19							1.0732	0.0826	0.0000	1.0000	0.7076
0.49	0.01							10.3725	0.1179	0.0000	1.0000	0.9785
Equation (1.8)[a]												
0.01	0.97	0	***	***	0	0.0014	***	0.0080	0.0030	0	0.9986	0.3880
0.10	0.79	0	0.0001	18.4139	0	0.0019	3.2382	0.0802	0.1187	0	0.9981	0.5319
0.20	0.59	0	0.0001	19.6642	0	0.0017	2.6567	0.2029	0.1005	0	0.9982	0.5830
0.30	0.39	0	0.0001	8.1486	0	0.0019	2.9423	0.4382	0.0570	0	0.9981	0.6274
0.40	0.19	0	0.0001	7.6018	0	0.0010	3.6334	1.0772	0.0824	0	0.9989	0.7054
0.49	0.01	0	0.0003	0.7745	0	0.0002	7.8400	10.3960	0.1178	0	0.9996	0.9770

Note *** denotes an estimate at the limit of the numerically calculated space–time data cube; C0 denotes the nugget, C1 the sill, and r the range parameter

[a]The three C0 terms were set equal to 0—the theoretical value—in order to achieve convergence

$(0.49, 0.01), (0.40, 0.19), (0.30, 0.39), (0.20, 0.59), (0.10, 0.79), (0.01, 0.97)\}$ (see Table 1.1). Theoretical nugget and sill values for (1.5)–(1.8) respectively are 0 and 1. Deviations from these values represent specification error; the numerical integration error is negligible.

The stable, the Bessel, and the exponential variogram models were evaluated in terms of their fits to these numerical data, with the exponential variogram model performing the best. Estimation results for this model appear in Tables 1.1 and 1.2, and suggest that (1.5) does not furnish a good description of the space–time structure generated by (1.1) and (1.2). Equation (1.8) fails to provide any improvement in the description furnished by (1.7), because (1.1) and (1.2) generate realizations from a random function that have the same sill (variance) in time and space; in cases where the variances differ, (1.8) will almost surely do better than (1.7). Equation (1.7) appears to yield a marginally better description than the one provided by (1.6). The principal difference between the relationship between (1.7) and (1.1) and (1.2) is the estimated α parameter, the anisotropic weight attached to the time distance in order to differentiate it from space distance, which is included in the specification of (1.7), but not (1.1) and (1.2).

1.5 Conclusions

In summary, numerical evaluation suggests that the STAR models (1.1) and (1.2) yield the metric model (1.7) with exponential-shaped variograms. However, real-world processes may, in addition to the space–time models characterized by (1.1) and (1.2), have purely spatial and purely temporal components. Whittle (1954) shows that purely spatial AR models have Bessel function-shaped covariance functions, whereas linear one-dimensional time series models have exponential variograms. Thus, processes that also have purely temporal and/or spatial components should be characterized by variogram models given by (1.8) rather than (1.7). Assuming that the generating process satisfies the linear ARMA model, the temporal and spatio-temporal variograms may be described with exponential functions, whereas the spatial component may be described with a Bessel function.

References

Bartlett M (1975) The statistical analysis of spatial pattern. Chapman-Hall, London

Basilevsky A (1983) Applied matrix algebra in the statistical sciences. North-Holland, New York

Bogaert P (1996) Comparison of kriging techniques in a space–time context. Math Geol 28:73–86

Brown P, Diggle P, Lord M, Young P (2001) Space–time calibration of radar rainfall data. Appl Stat 50:221–241

Calder C (2007) Dynamic factor process convolution models for multivariate space–time data with application to air quality assessment. Environ Ecol Stat 14:229–247

Cliff A, Haggett P, Ord J, Bassett K, Davies R (1975) Elements of spatial structure. Cambridge University Press, Cambridge

De Cesare L, Myers DE, Posa D (2001) Product-sum covariance for space–time modeling: an environmental application. Environmetrics 12:11–23

De Iaco S, Myers DE, Posa D (2003) The linear coregionalization model and the product-sum space–time variogram. Math Geol 35:25–38

Dimitrakopoulos R, Luo X (1994) Spatiotemporal modeling: covariances and ordinary kriging systems. In: Dimitrakopoulos R (ed) Geostatistics for the next century. Kluwer, Dordrecht, pp 88–93

Fuentes M, Chen L, Davis J (2008) A class of nonseparable and nonstationary spatial temporal covariance functions. Environmetrics 19:487–507

Gasim A (1988) First-order autoregressive models: a method for obtaining eigenvalues for weighting matrices. J Stat Plan Inference 18:391–398

Gething PW, Atkinson PM, Noor AM, Gikandi PW, Hay SI, Nixon MS (2007) A local space–time kriging approach applied to a national outpatient malaria data set. Comput Geosci 33:1337–1350

Gneiting T, Genton M, Guttorp P (2006). Chapter 4 – Statistical methods of Spatio-Temporal systems. In: Finkenstaedt B, Held L, Isham V (eds) Statistics of Spatio-temporal Systems. Chapman-Hall, Boca Raton, pp 151–175

Griffith D (1996) Spatial statistical analysis and GIS: exploring computational simplifications for estimating the neighborhood spatial forecasting model. In: Longley P, Batty M (eds) Spatial analysis: modelling in a GIS environment. Longman GeoInform, London, pp 255–268

Griffith D, Csillag F (1993) Exploring relationships between semi-variogram and spatial autoregressive models. Pap Reg Sci 72:283–295

Griffith D, Layne L (1997) Uncovering relationships between geo-statistical and spatial autoregressive models. In: 1996 proceedings on the section on statistics and the environment. American Statistical Association, Washington, DC, pp 91–96

Haining R (1978) Specification and estimation problems in models of spatial dependence. Department of Geography, Northwestern University, Evanston

Kolovos A, Christakos G, Hristopulos D, Serre M (2004) Methods for generating non-separable spatiotemporal covariance models with potential environmental applications. Adv Water Resour 27:815–830

Ma C (2003) Families of spatio-temporal stationary covariance models. J Stat Plan Inference 116:489–501

Ma C (2008) Recent developments on the construction of spatio-temporal covariance models. Stoc Environ Res Risk Assess 22:S39–S47

Mitchell M, Genton M, Gumpertz M (2005) Testing for separability of space–time covariances. Environmetrics 16:819–831

Ord J (1975) Estimation methods for models of spatial interaction. J Am Stat Assoc 70:120–126

Stein M (2005) Space–time covariance functions. J Am Stat Assoc 100:310–321

Whittle P (1954) On stationary processes in the plane. Biometrika 41:434–449

Part II
Spatial Analysis and Mining

Chapter 2
Sampling Survey of Heavy Metal in Soil Using SSSI

Aihua Ma, Jinfeng Wang, and Keli Zhang

Abstract Much attention has been given to sampling design, and the sampling method chosen, directly affects sampling accuracy. The development of spatial sampling theory has lead to the recognition of the importance of taking spatial dependency into account when sampling. This research study uses the new Sandwich Spatial Sampling and Inference (SSSI) software as a tool to compare the relative error, coefficient of variation (CV), and design effect of five sampling models: simple random sampling, stratified sampling, spatial random sampling, spatial stratified sampling, and sandwich spatial sampling. The five models are each simulated 1,000 times, with a range of sample sizes from 10 to 80. SSSI includes six models in all, but systematic sampling is not used in this study, because the sample positions are fixed. The dataset consists of 84 points measuring soil heavy metal content in Shanxi Province, China. The whole area is stratified into four layers by soil type, hierarchical cluster and geochronology, and three layers by geological surface. The research shows that the accuracy of spatial simple random sampling and spatial stratified sampling is better than simple random sampling and stratified sampling because the soil content is spatially continuous, and stratified models are more efficient than non-stratified models. Stratification by soil type yields higher accuracy than by geochronology in the case of smaller sample sizes, but lower accuracy in larger sample sizes. Based on spatial stratified sampling, sandwich

A.H. Ma (✉) • K.L. Zhang
School of Geography and Remote Sensing Sciences, Beijing Normal University, 100875 Beijing, China
e-mail: maaihua23@126.com; keli@bnu.edu.cn

J.F. Wang
Institute of Geographic Sciences and Nature Resources Research, Chinese Academy of Sciences, 100101 Beijing, China
e-mail: wangjf@lreis.ac.cn

A. Yeh et al. (eds.), *Advances in Spatial Data Handling and GIS*,
Lecture Notes in Geoinformation and Cartography,
DOI 10.1007/978-3-642-25926-5_2,

sampling develops a report layer composed of the user's final report units, allowing the user to obtain the mean and variance of each report unit with high accuracy. In the case of soil sampling, SSSI is a useful tool for evaluating the accuracy of different sampling techniques.

2.1 Introduction

Much progress has been made in developing tools to assess estimation accuracy and the relevance of chosen models, such as spatial autocorrelation precision (Cressie 1991; Wang et al. 2002; Haining 2003; Li et al. 2005) or error variance (Wang et al. 2010). Uncertainty in a spatial sampling survey can be reduced with high-quality information, improvement of statistical estimation accuracy, and increasingly precise geo-statistical software. The choice of a relevant sampling model, however, is essential to the quality of the analysis and that which is learned from the study.

Many different sampling methods exist, and selecting one, relevant to a specific problem, is always an important choice for the scientist, who wants to have the most valid statistical inference analysis possible. The user can decide to study the problem with, for instance, a simple random sampling model. This involves selecting a fixed number of units from a population, each with an equal chance, or the selection of a stratified sampling model to assess the relationship between samples and a stratified information layer (Cochran 1977). The method choice really depends on the problem, on the objective of the study, or on the available dataset. Having methods with which to assess the fit of the sample dataset with the sampling model would be useful to the scientist,and in all likelihood improve the statistical analysis and refine results.

In this study, two approaches for matching sampling effort to accuracy are chosen: a classical approach, which ignores spatial dependence between observations; and a geo-statistical approach, which accounts for spatial dependence (Hedger et al. 2001). Accounting for spatial dependency decreases the standard error when the environmental variable is spatially correlated (Wang et al. 2002; Haining 1988).

Sandwich Spatial Sampling and Inference (SSSI) is a professional spatial sampling and statistical inference tool. It can be used for environmental, resource, land, ecological, social, and economic sciences and practice sampling design and statistical inference. The objective of this study is to evaluate the accuracy of different sampling methods by comparing several indicators, such as relative error, coefficient of variation (CV), and design effect, using SSSI. Because the sampling model and number of samples (or sample size) are the two most important factors determining sampling efficiency (Chen et al. 2009), this study compares four models with a range of sample sizes from 10 to 80. The dataset is a survey of heavy metal content in soil in Shanxi province, China.

2.2 Material and Methods

2.2.1 Data and Site Description

The investigation and environmental sampling was carried out in 2002 and 2003. Two counties, Zhongyang and Jiaokou, in Shanxi Province were selected as the research area. There are seven administrative villages in each county and each has a high incidence of birth defects. Samples were collected using administration villages as units. Sample sites, consistent with birth defect epidemiological survey sites, were considered, so there exist some gaps within the data set, such as no samples in the western region. Soil samples were collected far from soil disturbance plots, such as those affected by residences and roads, to ensure, as far as possible, the true representation of local environmental characteristics. Samples were collected in most villages, resulting in a total of 84 points in all. Well-mixed cultivated land surface soil samples were acquired with a shovel 10–20 cm deep. The samples were dried, cleaned of organic debris and impurities, ground with a mortar, sifted with a 100 mesh aperture, and weighed at about 0.1 g soil 3 copies. They were then mixed with 3 ml of Glass Acid, 1 ml perchloric acid, and 1 ml hydrofluoric acid and placed into an oven, for 5 h, to bake and nitrify. Finally, they were put on an electric heating board for 1–2 h, mixed with water to give a uniform volume of 10 ml, for testing (Zhang et al. 2008). An ULTIMA inductively coupled plasma spectrometer was used to measure 16 elements in the soil: Al,As,Ca,Cu,Fe, K,Mg,Mo,Na,Ni,Pb,Se,Sn,Sr,V, and Zn. In this paper, Mo contents are used as sampling data.

Heavy metal content in the soil primarily depends on the natural levels of soil-forming parent materials and the absence of external contaminants (Wang et al. 1992). Through the process of weathering and soil formation, the concentration of these elements will change, but the relationship between parent material and their concentration in the soil is still relatively close. There are five main soil types in pedogenesis: cinnamon, chestnut cinnamon, loessial soil, regosols soil and Chao soil. In this way, the area is stratified by soil type and geological surface. It can also be stratified geochronologically, because elements can be introduced at different times. There are eight layers, but in order to keep the sample size reasonable in each layer, the region was stratified into four layers. The fourth stratification method is hierarchical clustering accomplished with SPSS.

2.2.2 Sampling Method and Algorithm

Five models in SSSI were used: simple random sampling, stratified sampling, spatial random sampling, spatial stratified sampling, and sandwich spatial sampling. The first two models took the classical approach, and the last ones took the geo-statistical approach. The use of different sampling designs and methods may

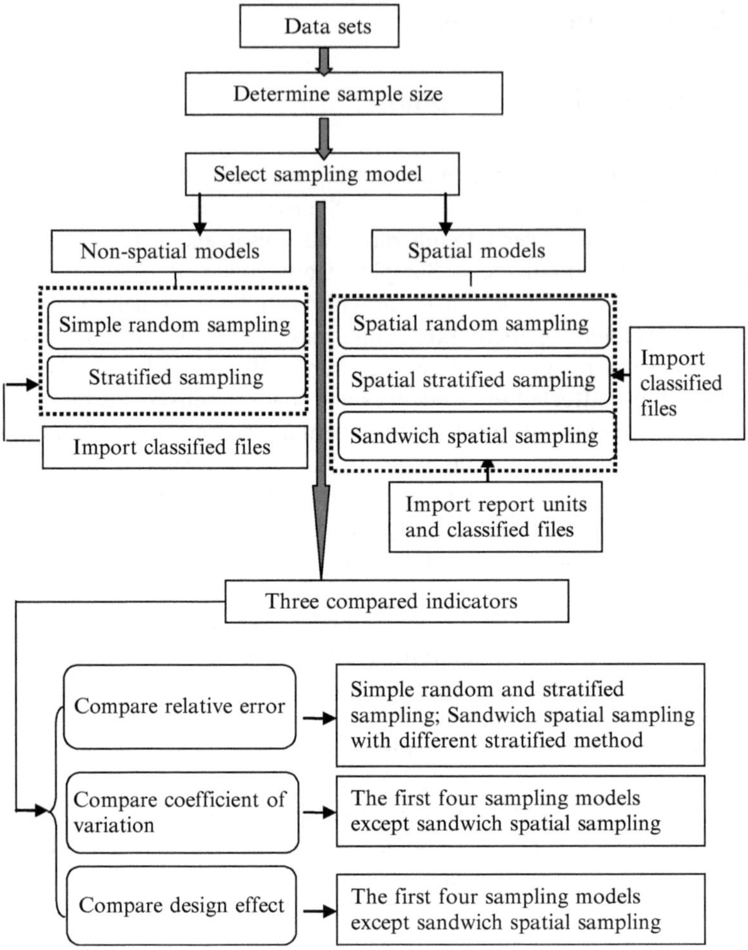

Fig. 2.1 A flow chart of the steps of this study

affect the accuracy of the analysis of samples collected in the same area (de Zorzi et al. 2008).In these cases, three indicators were selected to compare the efficiency of different sampling models. To select the three indicators, the models were simulated 1,000 times, each with a range of sample sizes from 10 to 80. To ease reader's understanding, Fig. 2.1 gives a flow chart of the steps taken:

2.2.2.1 Relative Error

Generally, the accuracy of sampling is higher with larger sample sizes, and the sample mean is nearer to the population mean. But at a certain sample size, the sampling method can affect the accuracy, so relative error (d_r) is a good indicator to judge which sampling model is more accurate. Relative error compares the difference between the sample mean and its true mean, so the estimated relative error is defined as:

$$d_r = \frac{abs(\overline{Y} - \overline{y})}{\overline{y}} \tag{2.1}$$

where $\overline{y}$ = sample mean

$\overline{Y}$ = observable population mean.

The former is estimated by the different models and the latter is acquired by counting the entire observable metal contents.

Figure 2.2 compares d_r of the simple random sampling and of the stratified sampling. The sandwich spatial sampling model is the newest method in the SSSI software. It has the same accuracy as spatial stratified sampling, but it considers the report layers, which can be in any unit, for example, a county border, provincial boundary, watershed, or artificial grid (Wang et al. 2009). Table 2.1 shows the values of the whole relative error of report layers.

2.2.2.2 Coefficient of Variation

The coefficient of variation is a statistical measure of the dispersion of data points in a data series around the mean. It is calculated as follows:

$$\text{coefficient of deviation} = \frac{\text{Standard deviation}}{\text{m}} \tag{2.2}$$

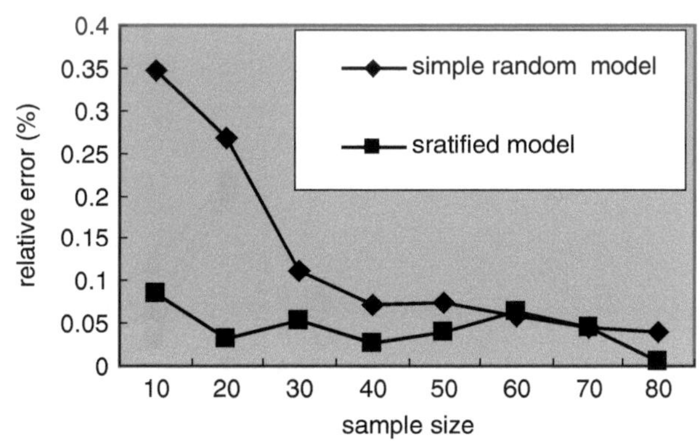

Fig. 2.2 Comparison of the relative error of simple random sampling and of stratified sampling

Table 2.1 Relative error of two kinds of report layer

Report layer	Stratified by soil type	Stratified by geochronology
Administrative villages	0.180	0.085
Grids	0.066	0.052

$$\text{Standard deviation} = \frac{\sum_{i=1}^{n} (x_i - \bar{x})^2}{n - 1} \qquad (2.3)$$

where m = sample mean
x_i = value of element content
n = sample size.

In this paper the dispersion of each estimator from its true value during the 1,000 simulations for each sampling method and sample size is compared (Chen et al. 2009).

Figures 2.3, 2.4, 2.5 and 2.6 show the CV from four sampling models of Mo elements, stratified by soil type, geological surface, hierarchical cluster, and geochronology. Figure 2.7 shows which stratified method is the most efficient.

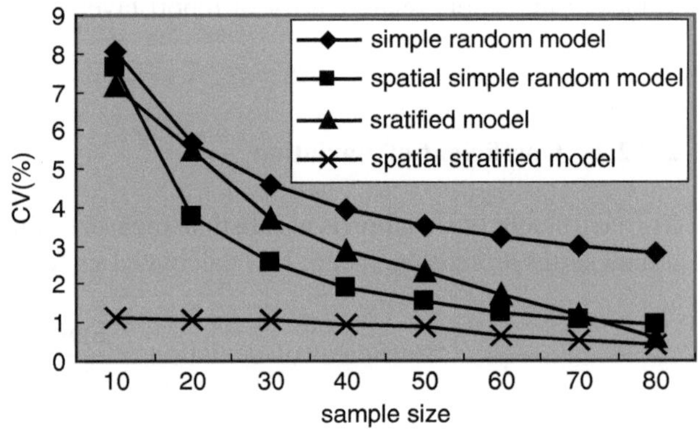

Fig. 2.3 Mo (stratified by soil type)

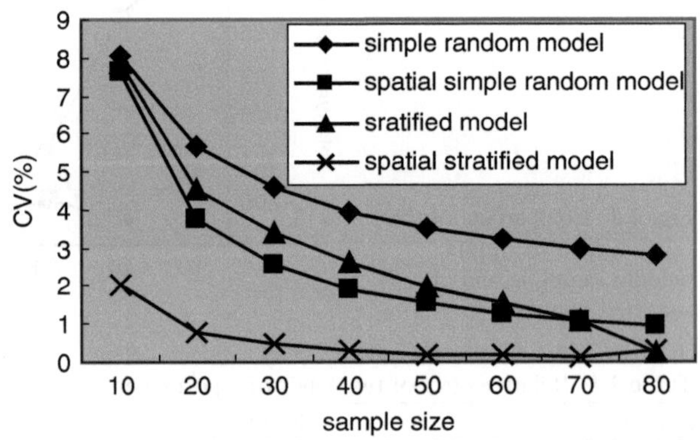

Fig. 2.4 Mo (stratified by geological surface)

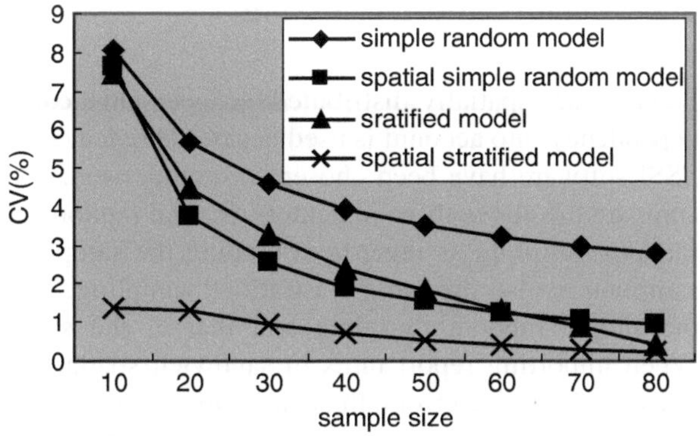

Fig. 2.5 Mo (stratified by Hierarchical cluster)

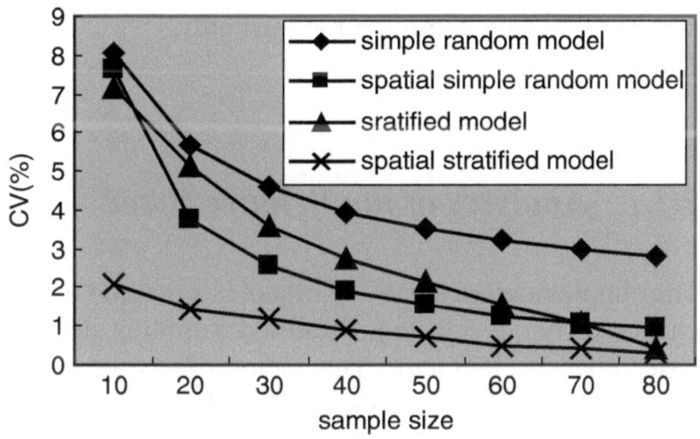

Fig. 2.6 Mo (stratified by geochronology)

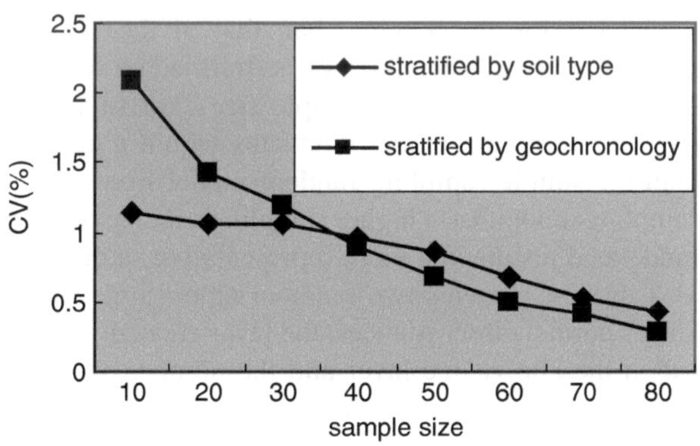

Fig. 2.7 Mo compare stratified method soil type and geochronology

2.2.2.3 Design Effect

The design effect is the ratio of variance of the estimate obtained from the (more complex) sample to the variance of the estimate obtained from a simple random sample of the same number of units (Cochran 1977). This indicator also enables the comparison of the efficiency of different sampling models.

2.3 Results and Discussion

Samples are spatially distributed, so geo-statistical approach which takes spatial dependency into account is used besides classical approach. Finally, five models of SSSI software have been chosen for comparison, while not all of them is used to compare the above three indicators. Spatial random sampling is the same as simple random sampling as regards evaluating the sampling mean, and spatial stratified sampling is also the same as stratified sampling. Hence only the relative error of the simple random sampling and that of the stratified sampling is compared. When importing report units in sandwich spatial sampling, each report unit has individual relative errors, thus all relative errors in the report layers are compared. Sandwich spatial sampling has the same accuracy as the spatial stratified sampling in coefficients of variation and design effect, hence these last two indicators are compared in four sampling models, only.

2.3.1 Analyses of the Relative Error

The simple random sampling model is a method in which each sample has an equal chance of being selected. Stratified sampling divides the whole area into groups based on a special factor. This factor affects the distribution of the research object; there is little variation inside the layer but great variation between layers. Figure 2.2 shows that as the sample size increases, the relative error in the simple random sampling model, decreases. With smaller sample sizes, the simple random sampling model has lower accuracy than that of the stratified sampling model, and the interval is large. When the area is stratified by soil type, each sample is represented in every layer, although sample sizes are small. With larger sample sizes, the stratified sampling model fluctuates within a certain range, but is more accurate than the simple sampling random model from 10 to 80. Therefore, the stratified sampling model has a higher sampling efficiency than the simple random sampling model and results in a more representative sample.

Table 2.1 presents two kinds of report units: administrative villages and grids. Under normal circumstances, the layer created with the stratified sampling model is called the knowledge layer and the number of knowledge layers is less than the number of report layers. The mean and variation of each report layer are obtained with SSSI, and using Formula (2.1), each report layer's relative error can been calculated. The whole relative error can be found by combining layers. The values of the table are less than 10%, except for administrative villages, stratified by soil type.

2.3.2 *Analyses of the Coefficient of Variance*

Figures 2.3, 2.4, 2.5 and 2.6 show the coefficient of variance (CV) of Mo for the simple random sampling model, spatial random sampling model, stratified sampling model, and spatial stratified sampling model. CV is the ratio of spatial variation to the sample mean. For small sample sizes, the estimated sample mean is less accurate than that for large sample sizes. The CV decreases with an increase in the sample size for all sampling models. But the curve fluctuation ranges are different for each sampling model. The spatial stratified sampling model has the highest precision, indicated by its curve, which is below those of the other graphs, and the simple random sampling model has the lowest precision. The spatial random sampling model has higher precision at the middle sample sizes at about from 15 to 70. It can be concluded that the heavy metal contents in soil have spatial autocorrelation and that the precision of the different sampling models is impacted more by the spatial dependency than by the areal heterogeneity for most sample sizes.

Taking spatial dependency into account, the nearer the samples are to one another, the more similar their attributes and the greater their spatial correlation. Spatial mean variation is calculated from the samples with the following formula:

$$\mathrm{Var}\{\bar{Z} - Z(A)/Z\} = \frac{1}{n}\left[\sigma^2 - E\{C(X,Y)\}\right] = \frac{\sigma^2}{n} - \frac{E\{C(X,Y)\}}{n} \tag{2.4}$$

where σ^2 = population variation
X, Y = random variables subject to uniform distribution in the area A
$C(X,Y)$ = the covariance of random variables X and Y.

Classical sampling mean variance is σ^2/n, and spatial sampling mean variance is the above formula with a decrease of $E\{C(X,Y)\}/n$ from classical sampling (Wang et al. 2009).

Spatial stratified sampling models take area heterogeneity and spatial dependency into account, not only to keep samples representative, but also to calculate sample autocorrelation. The sampling mean variance is smaller than any other sampling models, but the simple random sampling model does not account for area heterogeneity and spatial dependency, so the efficiency is lower. The stratified sampling and spatial random sampling models account for only area heterogeneity or spatial dependency, hence their efficiency is between spatial stratified sampling models and simple random sampling models. For small sample sizes, the stratified sampling model is better than the spatial random sampling model because it shows that the dependence of samples is not strong and the samples are more representative. With greater the sample size, the stronger the dependence . Although samples are more representative with the stratified sampling model, the impact degree is not larger than spatial dependence. When the sample size is about 60–70 (almost the population size) the spatial dependence reaches its maximum. It tends towards a certain level value, so there are some limitations for spatial dependence at small and large sample sizes.

As shown in Figs. 2.3, 2.4, 2.5 and 2.6, the accuracy of the simple random and spatial random sampling models is the same. A difference is shown in the stratified sampling model and spatial stratified sampling model because of the different stratification methods used. Selecting the best stratification method is important in the estimation of sampling precision. If the stratification is poor, the accuracy could be lower than that produced by simple random sampling. Figure 2.7 gives a comparison of the CV of two stratification methods. In the graph, stratification by soil type yields higher accuracy in the case of smaller sample sizes, than by geochronology, but lower accuracy in larger sample sizes. Knowledge layer is not necessarily spatially continuous, so samples are also not necessarily near each other even in the same layer. But the stratification of soil type is much more continuous than that of the geochronology type. We know that the closer the objects, the higher the similar degree of objects. When sampling a certain samples in each layer and the sample size is very small, a sample which obtains through stratification of soil type can be more representative, and it brings high sampling efficiency. In the larger sample sizes, however, geochronological stratification is better, possibly because the Mo content has been more strongly affected by geochronology. This effect is manifested,when the sample size reaches a certain degree.

2.3.3 Analyses of the Design Effect

Tables 2.2 and 2.3 show the ratio of mean variance for different sampling models to the mean variance from the simple random sampling model. The values in the tables are all less than one, indicating that the simple random sampling model (Srs) is the least efficient and the spatial stratified sampling model (SStrs) is the most efficient. Spatial autocorrelation slightly impacted the sampling accuracy.

The spatial stratified sampling model produces accuracy, higher than any other sampling models. In traditional stratified sampling method, objects within the same

Table 2.2 Design effect of spatial simple sampling, (spatial) stratified sampling by soil type and geological surface

	Models				
Sample sizes	Srs	StrRs (a)	SStrs (a)	StrRS (b)	SStrs (b)
10	0.945	0.891	0.143	0.899	0.228
20	0.664	0.964	0.187	0.798	0.133
30	0.557	0.801	0.230	0.745	0.100
40	0.485	0.719	0.241	0.661	0.070
50	0.430	0.664	0.244	0.556	0.059
60	0.392	0.536	0.209	0.469	0.048
70	0.361	0.402	0.177	0.351	0.043
80	0.338	0.203	0.152	0.092	0.096

Srs spatial random sampling, *SStrs* spatial stratified sampling model
a stratified by soil type, *b* stratified by geological surface

Table 2.3 Design effect of (spatial) stratified sampling by hierarchical cluster and geochronology

	Models			
Sample sizes	StrRs (c)	SStrs (c)	StrRS (d)	SStrs (d)
10	0.931	0.172	0.889	0.261
20	0.788	0.228	0.901	0.252
30	0.710	0.203	0.776	0.281
40	0.606	0.178	0.692	0.225
50	0.523	0.156	0.612	0.193
60	0.405	0.127	0.474	0.150
70	0.293	0.106	0.359	0.137
80	0.150	0.087	0.155	0.101

SStrs spatial stratified sampling model
c stratified by hierarchical cluster, *d* stratified by geochronology

knowledge layer may be very far apart and even spatially separated from the other layers. Spatial stratified sampling not only makes sure that variation is small in the same layer and large between layers, but also requires the layer is spatially continuous. In these four stratification methods, the second meet the requirement of keeping objects in the same layer distributed spatially continuous. So from Tables 2.2 and 2.3, it is seen that b has lower values for most sample sizes except at 10 and 80, and c has higher accuracy at samples,10 and 80. Samples in the same knowledge layer in space have high spatial dependence, and $E\{C(X,Y)\}/n$ also increases. But at 10 and 80, the spatial dependence limitations become apparent. Hierarchical clustering stratifies samples by their Euclidean distance squared, so the properties of each layer are consistent within the sample, when compared to the results of other stratification methods, but samples are not together in space, resulting in high sampling efficiency outside the limitations of space.

2.4 Conclusions

In this research study, both spatial and non-spatial models have been investigated, with comparisons being made between five types of sampling models. Of these, sandwich spatial sampling and spatial stratified sampling were shown to have the highest efficiency, while the simple random model had the lowest efficiency. The efficiency of spatial random sampling and stratified sampling is shown to be dependent on the sample size. The stratified sampling model provides higher efficiency than spatial random sampling at both large and small sample sizes, but has lower efficiency at middle sizes or sizes between the two extremes. The sampling efficiency is also much dependent on the stratification method. As has been indicated, different methods can affect the sample distribution. For example, when stratified by different stratification method, even if two samples which are adjacent in space may be in different layers. The sample efficiency would be higher when the stratified factors can reflect the research objects' real spatial variance better. Here, it means that the factor whose spatial distribution is closer to the real distribution of soil element content should be selected as knowledge stratification. At the same sample size. Spatial stratified sampling takes spatial dependence into account and keeps samples, together, in the same spacial knowledge layer, so stratification by geological surface has higher efficiency than other stratification methods. By comparing two kinds of stratification methods, by soil type and geochronology, it can be seen that, in the case of smaller sample sizes, stratification by soil type yields higher accuracy than by geochronology, but yields lower accuracy in larger sample sizes. Sandwich spatial sampling has the same efficiency as spatial stratified sampling for the sample mean and spatial mean variation. These values, however, can be calculated for each report layer. These report layers can then be divided arbitrarily, to meet the user's needs.

Acknowledgement The authors would like to thank Professor Chaoyang Wei and Tongbin Chen for their helpful discussions and constructive suggestions, and also appreciate the assistance of Luke ldrisk, Antoine Soullié and Hexiang Ba in the programming.

References

Cochran WG (1977) Sampling techniques, 3rd edn. Wiley, New York, pp 85–88

Cressie N (1991) Statistics for spatial data. Wiley, New York

de Zorzi P, Barbizzi S, Belli M, Mufato R, Sartori G, Stocchero G (2008) Soil sampling strategies: evaluation of different approaches. Appl Radiat Isot 66:1691–1694

Chen DM, Wei H (2009) The effect of spatial autocorrelation and class proportion on the accuracy. ISPRS J Photogramm Remote Sens 64:140–150

Wang GY, Pan M, Liu XD, Liang HH (1992) On the relationship between the concentrations of elements in soil and the types of soil-forming parental material in Shandong province, China. Acta Sci Nat Univ Pekin 28(4):475–485

Haining RP (1988) Estimating spatial means with an application to remote sensing data. Commun Stat Theory Methodol 17(2):537–597

Haining RP (2003) Spatial data analysis: theory and practice. Cambridge University Press, Cambridge

Hedger RD, Atkinson PM, Malthus TJ (2001) Optimizing sampling strategies for estimating mean water quality in lakes using geostatistical techniques with remote sensing. Lakes and Reservoirs: Research and Management 6(4):279–288

Wang JF, Liu JY, Zhuang DF, Li LF, Ge Y (2002) Spatial sampling design for monitoring the area of cultivated land. Int J Remote Sens 23(2):263–284

Wang JF, Haining RP, Cao ZD (2010) Sample surveying to estimate the mean of a heterogeneous surface: reducing the error variance through zoning. Int J Geogr Inf Sci 24(4):523–543 doi:10.1080/13658810902873512

Wang JF, Jiang CS, Li LF, Hu MG (2009) Spatial sampling and statistical inference (in Chinese). Science Press, Beijing, pp 17–20

Li LF, Wang JF, Liu JY (2005) Optimal decision-making model of spatial sampling for survey of Chinas land with remotely sensed data. Sci China Ser D 48(6):752–764

Zhang ZD, Zhang KL, Wu JL, Zhang T, Zheng XY (2008) Spatial information analysis of characteristics of soil trace element contents in high incidence areas of birth defects. J Zhejiang Univ (Agric1 Life Sci1) 34(6):684–691

Chapter 3
An Optimised Cellular Automata Model Based on Adaptive Genetic Algorithm for Urban Growth Simulation

Yongjiu Feng and Yan Liu

Abstract This paper presents an improved cellular automata (CA) model optimised using an adaptive genetic algorithm (AGA) to simulate the spatio-temporal processes of urban growth. The AGA technique was used to optimise the transition rules of the CA model defined through conventional logistic regression approach, resulting in higher simulation efficiency and improved results. Application of the AGA based CA model in Shanghai's Jiading District, Eastern China demonstrates that the model was able to generate reasonable representation of urban growth even with limited input data in defining its transition rules. The research shows that AGA technique can be integrated within a conventional CA based urban simulation model to improve human understanding on urban dynamics.

3.1 Introduction

This paper presents a method applying an adaptive genetic algorithm (AGA) to define and search for transition rules and parameters of a cellular automata (CA) model to simulate the spatio-temporal processes of urban growth. This method builds on traditional CA approach and applies an intelligent search algorithm to optimise the configuration of a set of spatial parameters, resulting in higher simulation efficiency and improved simulation outcomes.

Y.J. Feng
College of Marine Sciences, Shanghai Ocean University, 999 Huchenghuan Road, Shanghai, People's Republic of China
e-mail: yjfeng@shou.edu.cn

Y. Liu (✉)
Centre for Spatial Environmental Research and School of Geography, Planning and Environmental Management, The University of Queensland, Australia, St Lucia, QLD, Australia
e-mail: yan.liu@uq.edu.au

A. Yeh et al. (eds.), *Advances in Spatial Data Handling and GIS*,
Lecture Notes in Geoinformation and Cartography,
DOI 10.1007/978-3-642-25926-5_3,

Conventionally, CA based urban models require strict definition of various spatial and non-spatial factors driving the development of urban growth (Batty and Xie 1994; Batty et al. 1999; Couclelis 1997; He et al. 2006; Li and Yeh 2002a, b; Liu 2008; Liu and Phinn 2003; Stevens and Dragićević 2007; Stevens et al. 2007; White and Engelen 1993; Wu 1998, 2002). Many CA models have been developed using a diverse range of methods to define such factors and configure the relevant parameters; these methods include multi-criteria evaluation, logistic regression, principal component analysis, and partial least squares regression, to name a few. However, limitations of such methods in defining suitable transition rules, or the values of relevant parameters of the transition rules, or in constructing the architecture of the models have been identified and reported in the literature (Al-kheder et al. 2008; Li and Yeh 2002a). As a result, there are significant differences between the simulation results and the actual patterns of urban growth, making such models less accurate in simulating the actual process of urban growth (Li and Yeh 2002b; Liu and Phinn 2003).

The development of the adaptive genetic algorithm method has provided researchers with new ways to identify and search for suitable transition rules and their defining parameters in urban modelling (Bies et al. 2006; Li et al. 2007; Srinivas and Patnaik 1994). A genetic algorithm can be used to search for an optimal solution of a problem based on the mechanics of natural genetics and natural selection (Schmitt et al. 1998). This technique has been applied to address various optimization problems, such as satellite imagery classification (Huang et al. 2007) and problems clustering (Lorena and Furtado 2001).

The following section presents a generic CA model based on logistic regression method first, followed by the adaptive genetic algorithm used to optimise the CA parameters, minimising the differences between the simulated results and the actual urban development. The next section applies the AGA based CA model to simulate the urban growth of Shanghai's Jiading District, Eastern China. Results from the model are also presented and discussed in this section, followed by conclusions in the last section.

3.2 The Adaptive Genetic Algorithm Based CA Model

3.2.1 A Generic CA Model Based on Logistic Regression

Generally, CA defines the state of a cell at one time as a function of the state of the cell and states of its neighbouring cells at a previous time in accordance with a set of transition rules, which can be generalised as follows (Wu 1998):

$$S_{ij}^{t+1} = f(S_{ij}^{t}, \Omega_{ij}, con()) \tag{3.1}$$

where $S_{ij}{}^{t}$ and $S_{ij}{}^{t+1}$represent the states of a cell at location *ij* at time *t* and *t* + 1, respectively; Ω_{ij} is a neighbourhood function; $con()$ defines a set of constraint factors affecting the transition of cell states; *f* is a transition function.

Assume that a cell can take one of only two states, urban or non-urban. A rectangular neighbourhood is defined with u × u cells; all cells within the neighbourhood have equal opportunity for development. Thus, the probability a cell changes its state from non-urban to urban can be defined as:

$$P^{t}_{\Omega} = \frac{\sum_{u \times u} (S^{t}_{ij} = Urban)}{u \times u - 1} \tag{3.2}$$

where P^{t}_{Ω} is the probability a cell can change from one state to another at time *t*; $\sum_{u \times u} (S^{t}_{ij} = Urban)$ represents the total number of urban cells at time *t* within the u × u cells neighbourhood.

However, in practice, not all cells have equal opportunity for development. For instance, some non-urban areas such as large-scale water bodies or areas with critical physical constraints such as very steep slope may not be able to develop into urban areas. Other areas such as the natural conservation and primary farmland may be prevented from urban development through institutional control, i.e., land use planning regulation.

In addition, a stochastic factor can be introduced into the CA based urban model to represent the unequal opportunity of development amongst all cells (Wu 2002). With a stochastic control factor, the probability a cell converts from non-urban to urban state can be defined as:

$$P^{t}_{G} = \frac{1}{1+\exp(-(a_1x_1+\ldots+a_mx_m))} \times P^{t}_{\Omega} \times con(cell^{t}_{ij} = suitable) \times (1 + (-\ln \gamma)^{\beta}) \tag{3.3}$$

where

P^{t}_{G} is the probability of the cell converting from one state to another at time *t;*

$con(cell^{t}_{ij} = suitable)$ is a constraint function, the value of which ranges from 0 to 1, with 0 meaning that the cell is constrained from changing its current state, and 1 meaning that the cell is able to change its state at the following time step;

$1 + (-\ln \gamma)^{\beta}$represents a stochastic factor, where γ is a random real number ranging from 0 to 1, and β is a parameter controlling the effect of the stochastic factor. The value of β ranges from 0 to 10;

x_i $(i = 1, 2, \ldots, m)$ are a series of spatial driving factors to urban growth, which can be represented by the distances from a cell to urban centres, town centres, main roads, and so on. These distance factors are also called spatial variables or independent variables; and

$a_1, a_2, \ldots, a_m$ are used to assign weight to each of the distance variables.

One of the common challenges in developing a CA model based on logistic regression method is how to choose the distance variables and configure the relevant parameters defining the impact of such distance factors on urban growth. Consequently, results generated by a logistic based CA model may show poor match for the actual patterns of urban growth. This indicates that there is a need to search for other techniques in identifying and defining the model's transition rules. Such a challenge in model development can be addressed by incorporating the adaptive genetic algorithm approach to randomly search for an optimised conversion probability for each cell, and subsequently minimise the differences between simulated results and actual urban growth patterns.

3.2.2 Adaptive Genetic Algorithm Based CA (AGA-CA) Modelling

A genetic algorithm is a search technique used in computing to find solutions to optimisation and search problems (Schmitt et al. 1998). Inspired by evolutionary biology, genetic algorithms work in computer simulations to search for an exact or approximate solution from a population of solutions (Liao et al. 2010). This search and optimisation process is achieved according to natural selection, including inheritance, selection, crossover and mutation.

There are two elements of a genetic algorithm, including a genetic representation of the solution domain, such as an array of cells in a cellular urban space, and a fitness function to evaluate and quantify the optimality of a solution.

The efficiency of a standard GA depends largely on the setting of its parameters such as the selection, crossover and mutation rates, which are difficult to adjust manually. Such difficulties can be overcome by the adaptive genetic algorithm as the AGA could dynamically modify the parameters of the genetic algorithm (Espinoza et al. 2001; Kee et al. 2001). AGAs not only keep population diversity effectively but also improve the performance of the local and premature convergences. Such genetic diversity is important to ensure the existence of all possible solutions in the solution domain and the identification of optimised solution. In addition, the adaptive genetic algorithm also enhances the search speed and precision of the genetic algorithm. Hence, the searching and optimisation process for problem solutions can be accelerated.

Fitness function A fitness function is an objective function to quantify the optimality of a solution. This function was created by selecting sample of cells within the cellular urban space to minimise the differences between the simulation results produced by a logistic regression based CA model and the actual urban growth patterns identified from remotely sensed images. The fitness function is defined and optimised through the modelling process as:

$$f(x) = \sum_{i=1}^{n} \left(f_i(x_1, x_2, ..., x_m) - f_i^0(x_1, x_2, ..., x_m)\right)^2 \tag{3.4}$$

where

$f(x)$ is a fitness function;

n is the number of samples selected from the cellular space which were used to retrieve the CA transition rules;

f_i is the conversion probability of the state of cell i based on the logistic regression model, i.e., $f_i = P_G$ as defined in (3.3); and

f_i^0 is the actual conversion decision of cell i. f_i^0 can only take one of the two values, 0 or 1, with $f_i^0 = 0$ meaning the state of the cell i remains as non-urban and $f_i^0 = 1$ meaning the state of the cell has changed from non-urban to urban.

The process of urban growth can be affected by many factors, including socio-economic, physical and environmental as well as institutional control factors. These factors can be built into the cellular model through a set of transition rules. With the fitness function, the simulation process of urban growth can be calibrated by dynamically updating the various parameters of the transition rules to minimise the value of the fitness function so the simulated urban patterns can better match for the observed patterns of urban growth. The model calibration process is completed once the fitness function reaches a stable value over time and the model's transition rules and parameters can be considered suitable for operation to the entire cellular space.

Coding of chromosomes Chromosomes are the abstract representation of candidate solutions, or individuals. A chromosome is a set of parameters which defines a proposed solution to the problem that the AGA is trying to solve. In the CA based urban modelling practice, all possible CA transition rules affecting urban growth are considered as chromosomes. Each chromosome is coded as a simple string like:

$$C = [\alpha_1^k, \alpha_2^k, \ldots, \alpha_m^k] \tag{3.5}$$

where C represents a string of candidate solutions; m is the number of spatial driving factors (as in Equation (3.3)); k means the kth individual (candidate solution). α_1^k to α_m^k represent the weight of each spatial driving factor in the kth candidate solution. The values of α_1^k to α_m^k in the optimised candidate solution are the parameter values required by the CA model as defined in Equation (3.3).

Initially a number of chromosomes were randomly generated to form the possible solutions for the adaptive genetic algorithm to begin its searching process. After many generations of selection, crossover and mutation operations, only those chromosomes which acquire lower fitness values will remain, resulting in the emergence of a good chromosome structure.

Selection operator Selection is the key operation of the AGA method in which individual genomes are chosen from a population of candidate solutions for later breeding, including recombination and crossover. During each successive generation, individual solutions are selected through a fitness-based process, where solutions with lower fitness values are typically more likely to be selected. Using the Hamming distance that measures the minimum number of substitutions required to change one string into the other as a selection criterion, one chromosome is selected from every randomly selected pair of chromosomes on a competitive selection process. The selection process ensures that the diversity of chromosomes is reserved during the selection process.

Crossover operator Crossover is an exchange of genetic material between homologous individuals for final genetic recombination. While many crossover operators are available in the genetic algorithms, this research employs the adaptive genetic operator proposed by Srinivas and Patnaik (1994). The crossover probability P_c is used to allow the crossover between chromosomes. This probability value changes continuously with the change of fitness value during the search process. The crossover probability is defined as:

$$P_c = \begin{cases} P_{c1} - \frac{(P_{c1}-P_{c2})(f'-f_{avg})}{f_{\max}-f_{avg}} & f' \geq f_{avg} \\ P_{c1} & f' < f_{avg} \end{cases} \tag{3.6}$$

where

P_{c1} is the maximum crossover probability, which is defined as 0.95;
P_{c2} is the minimum crossover probability, which is defined as 0.45;
f' is the fitness value;
$f_{\max}$ is the maximum fitness value; and
f_{avg} is the minimum fitness value.

Mutation operator In genetic algorithms, mutation is used to maintain genetic diversity from one generation of a population of chromosomes to the next. Similar to the crossover operator, the mutation operator proposed by Srinivas and Patnaik (1994) was adopted in this research. The mutation probability P_m is defined as:

$$P_m = \begin{cases} P_{m1} - \frac{(P_{m1}-P_{m2})(f_{\max}-f_{avg})}{f_{\max}-f_{avg}} & f' \geq f_{avg} \\ P_{m1} & f' < f_{avg} \end{cases} \tag{3.7}$$

where

P_{m1} is the maximum mutation probability, which is given as 0.1; and
P_{m2} is the minimum mutation probability, which is given as 0.001.

Threshold of conversion probability The AGA technique is used to minimise the fitness function $f(x)$ corresponding to the selected spatial samples. With the minimum value of $f(x)$, a set of optimised chromosome can be achieved together with its defining parameters. This leads to the generation of the conversion probability of each cell from non-urban to urban in the urban growth process. Hence, by comparing the conversion probability of a cell at time t with a pre-defined threshold value (Wu 1998; Wu 2002; Li and Yeh 2002b), if the conversion probability of the cell at time t is larger than the pre-defined threshold value the cell will be converted to an urban state at the following step. Otherwise, the state of the cell will remain unchanged.

$$S_{ij}^{t+1} = \begin{cases} Urban, & if\ P_G^t > P_{\text{threshold}} \\ Non\text{-}urban, & if\ P_G^t \leq P_{\text{threshold}} \end{cases} \tag{3.8}$$

3.3 Application

3.3.1 Study Area and Data

The proposed AGA-CA model was applied to simulate the urban growth in Shanghai's Jiading District, which is located in the Yangtze River Delta of Eastern China. The study area consists of 11 blocks (towns) with a total area size of 463.6 km^2. Rapid urban expansion had occurred in the 1990s due to the fast economic development and population growth. Urban growth of this area from 1989 to 2006 was mapped out using data from various sources, including two Landsat-5 Thematic Mapper (TM) images acquired on August 6, 1989 and April 30, 2006 respectively to obtain spectrum information of the study area. In addition, other essential ancillary data include a 1:50,000 topographic map, some cadastre and transportation maps collected from relevant local government agencies.

3.3.2 Model Configuration and Implementation

While many spatial factors can make an impact on urban growth, in practice, not all factors can be quantified into a simulation model, especially when data reflecting such factors are either not available or not accessible. Historically, the urban growth of Jiading District has been largely related to the existing distribution of urban and towns, the accessibility to transport as well as the preservation of primary agriculture land. Therefore, five spatial factors were selected, including distances to urban centre (x_1), town centre (x_2), main roads (x_3), cropland (x_4), and orchard field (x_5). The impact of each factor on urban development may be different, hence, different weights were assigned to each of these factors represented by a_1, a_2, a_3, a_4, a_5, respectively.

A total of 5,000 sample cells were selected randomly from the Landsat TM images for the AGA model to commence the search process. The distances from each sample cell to the urban centre, town centre, main road, cropland and orchard field were computed in GIS. These distance values were normalised to receive a standardised value ranging from 0 to 1.

A modelling framework was developed within ESRI's ArcGIS environment based on Microsoft Visual Basic.NET and ArcGIS Engine 9.2 technologies. This modelling framework incorporates the AGA-CA model as well as a number of other CA based modelling approaches. The user-friendly graphical user interface makes ease the sophisticated computation process of the model (Fig. 3.1). Moreover, the strong coupling of the model within a GIS framework makes it possible to use the various display and analysis functions of GIS in raster based data integration and modelling. Hence, this modelling framework becomes an important component of the AGA-CA program for modelling urban growth.

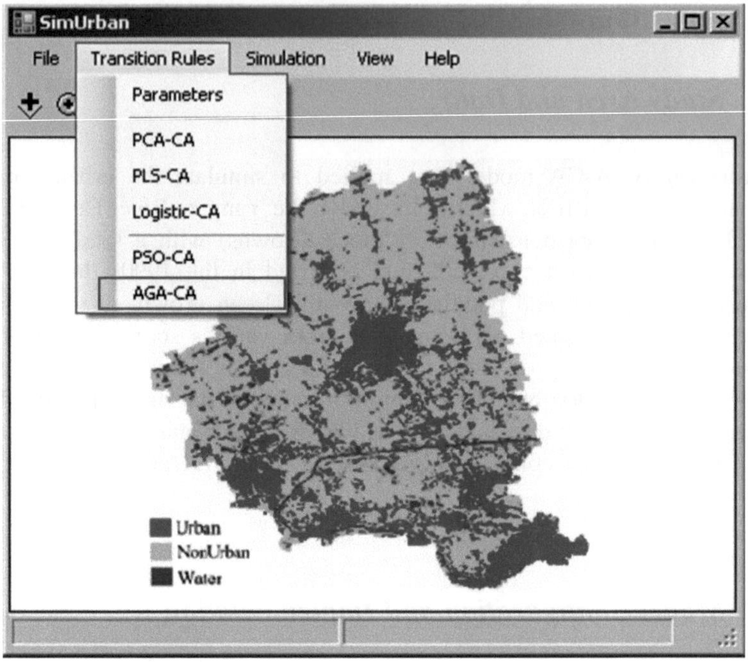

Fig. 3.1 GUI of the modelling framework

3.4 Results and Discussion

The AGA-CA model was launched to start the search and optimisation process with the sample data selected from the 1989 Landsat TM imagery. The fitness track in the evolutionary computation process demonstrates a rapid convergence rate after over 30,000 times of iteration, with a convergence value of 391.9855 (Fig. 3.2).

The convergence of the fitness track leads to the identification of a set of optimised chromosome or solution as:

$$C = \begin{cases} a_1 = -0.5083 \\ a_2 = -0.6417 \\ a_3 = -0.4962 \\ a_4 = +0.3723 \\ a_5 = +0.2174 \end{cases} \tag{3.9}$$

This optimised solution becomes the initial input data for the CA model to compute the conversion probability of cells from non-urban to urban state. Hence, Equation (3.3) can be re-written as:

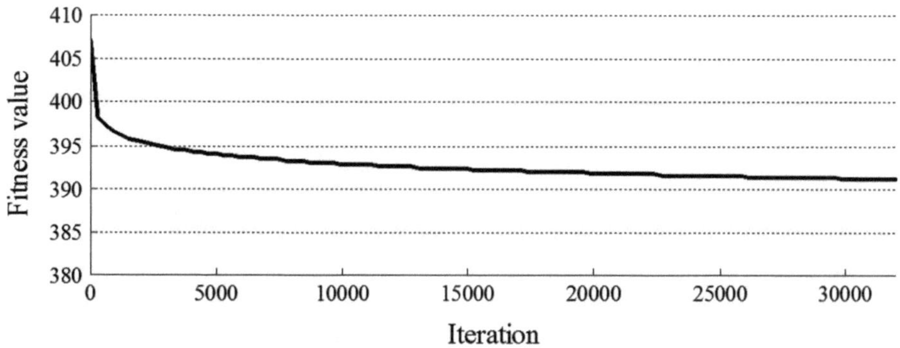

Fig. 3.2 Fitness track of the AGA model

$$P_G = [1 + \exp(-(-0.5083x_1 - 0.6417x_2 - 0.4962x_3 + 0.3723x_4 + 0.2174x_5))]^{-1} \times P_\Omega \times con(cell^t_{ij} = suitable) \times (1 + (-\ln\gamma)^\beta) \quad (3.10)$$

The optimised chromosome displayed in Equation (3.9) shows different effect of the spatial factors on urban growth in Shanghai's Jiading District. According to Equation (3.9), a negative $a_i (i = 1, ..., 5)$ leads to a larger P_G value, i.e., a higher possibility for a cell to convert from non-urban to urban state. Likewise, a positive a_i results in a lower P_G value, hence, a lower possibility for the cell to be converted into an urban state at the next time step. The optimised results show that the distance to town centres has the most significant impact on the development of cells within its neighbourhood. This is reflected by the smallest value of a_2 ($a_2 = -0.6417$). Likewise, the spatial proximities of a cell to urban centre and main road also play positive roles to its urban development (with $a_1 = -0.5083$ and $a_3 = -0.4962$ respectively). On the other hand, factors such as distances to cropland and orchard field have negative impact on urban development (with $a_4 = 0.3723$ and $a_5 = 0.2174$). Hence, the closer a cell is to cropland and orchard fields, the lower probability the cell has to develop into an urban state. This is a reflection of the conservation policy for primary agricultural land in practice.

Using the urban distribution patterns defined from the 1989 satellite image classification as the initial input data for the urban CA model (Fig. 3.3a), and the transition rules generated by the AGA approach, the CA model was operated to generate a series of urban growth scenarios. Each iteration of the model represents one year. After 16 iterations the model generates a map representing urban growth patterns of Jiading District in 2006 (Fig. 3.3c). This simulated urban scenario was compared with the actual urban distribution as defined by the land use classification from the 2006 Landsat TM image (Fig. 3.3b).

There is a high similarity between Figure 3.3b and c shows high similarity even by visual inspection and comparison. Comparing the two maps on a cell-by-cell basis, an error matrix analysis was realised (Table 3.1). The results show that the producer's accuracy for non-urban and urban areas were 85.7% and 76.3%

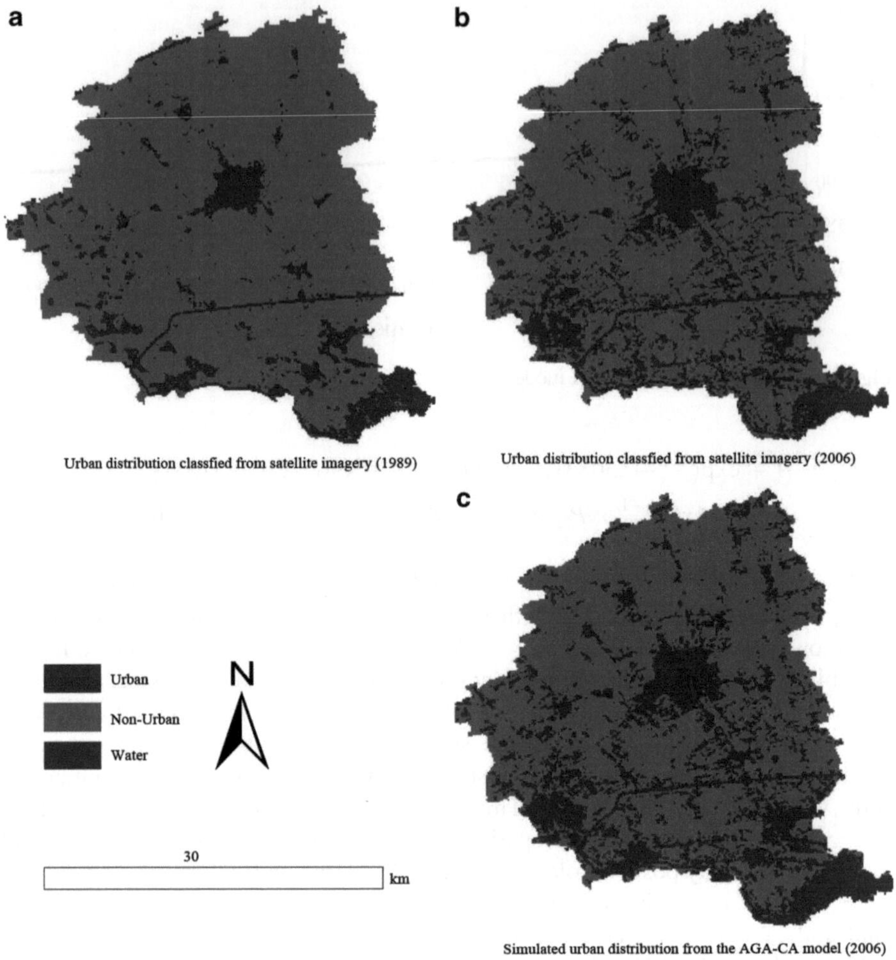

Fig. 3.3 Actual and simulated urban scenarios of Jiading District using the AGA-CA model

Table 3.1 The confusion matrix between remote sensing-based land use classification and the simulated urban categories using the AGA-CA model of Jiading District in 2006

	Simulation results (no. of cells)	
	Non-urban	Urban
Satellite-based land use classification (no. of cells)		
Non-urban	31,556	5,261
Urban	4,115	13,211
Producer's accuracies (%)		
Non-urban	85.7	
Urban	76.3	
User's accuracies (%)		
Non-urban	81.1	
Urban	86.7	
Overall accuracy (%)	82.7	
Kappa coefficient (%)	60.9	

respectively, while the user's accuracy for the non-urban and urban categories were 81.1% and 86.7%, respectively. Consequently, the model generated an overall accuracy of 82.7% and a Kappa coefficient of 60.9%. These simulation accuracies were considered good given that only five spatial distance variables were considered in the model. Should other factors such as the social demographic controls, institutional policy effects concerning sustainable urban development as well as other economic constraints included into the model, the AGA-CA model would be able to generate more accurate results.

3.5 Conclusions

This paper presents an improved CA model optimised by adaptive genetic algorithm technique, which has been developed as an evolutionary computation technique. By using the adaptive genetic algorithm, a set of transition rules and their defining parameters have been identified and optimised using the limited data available as input data sources. The AGA technique is particularly useful in optimising the CA transition rules which can be used by conventional CA model based on logistic regression approach. The application of the AGA-CA model in Shanghai's Jiading District demonstrates the effectiveness of the AGA technique in transition rule optimisation for CA based urban models, which contributes positively to human understanding on urban dynamics.

Further research will focus on introducing other spatial factors such as differentiated land values, social demographic controls and institutional policy effects into the modelling process. This will evaluate whether the adaptive genetic algorithm is capable of searching for and extracting more complex transition rules on urban land use change dynamics. In addition, it is necessary to reapply and test the applicability of the model in other regions such as the rapidly growing urban region in Southeast Queensland, Australia.

Acknowledgment This study was supported by the Innovation Program of Shanghai Municipal Education Commission (project no. 11YZ154), the Special Research Fund for Selected Outstanding Young University Scholars in Shanghai (project no. SSC09018), and the University of Queensland New Staff Research Start-up Fund (project no. 601871).

References

Al-kheder S, Wang J, Shan J (2008) Fuzzy inference guided cellular automata urban-growth modelling using multi-temporal satellite images. Int J Geogr Inf Sci 22(11):1271–1293

Batty M, Xie Y (1994) From cells to cities. Environ PlannB Plann Des 21:s31–s48

Batty M, Xie Y, Sun Z (1999) Modelling urban dynamics through GIS-based cellular automata. Comput Environ Urban Syst 23(3):205–233

Bies RR, Muldoon MF, Pollock BG, Manuck S, Smith G, Sale ME (2006) A genetic algorithm-based, hybrid machine learning approach to model selection. J Pharmacokinet Pharmacodyn 33(2):196–221
Couclelis H (1997) From cellular automata to urban models: new principles for model development and implementation. Environ PlannB Plann Des 24(2):165–174
Espinoza F, Minsker BS, Goldberg D (2001) A self-adaptive hybrid genetic algorithm. In Spector L, Goodman E, Wu A, Langdon WB, Voigt H-M, Gen M, Sen S, Dorigo M, Pezeshk S, Garzon M, Burke E (Eds.) GECCO 2001: Proceedings of the Genetic and Evolutionary Computation Conference. Morgan Kaufmann Publishers, San Francisco
He CY, Okada N, Zhang QF, Shi PJ, Zhang JS (2006) Modelling urban expansion scenarios by coupling cellular automata model and system dynamic model in Beijing, China. Appl Geogr 26 (3–4):323–345
Huang MX, Gong JH, Zhou S, Liu CB, Zhang LH (2007) Genetic algorithm-based decision tree classifier for remote sensing mapping with SPOT-5 data in the Hongshimao watershed of the loess plateau, China. Neural Comput Appl 6(6):513–517
Kee E, Airey S, Cye W (2001) An adaptive genetic algorithm. In Spector L, Goodman E, Wu A, Langdon WB, Voigt H-M, Gen M, Sen S, Dorigo M, Pezeshk S, Garzon M, Burke E (Eds.) GECCO 2001: Proceedings of the Genetic and Evolutionary Computation Conference. Morgan Kaufmann Publishers, San Francisco pp 391–397
Li X, Yeh AGO (2002a) Neural-network-based cellular automata for simulating multiple land use changes using GIS. Int J Geogr Inf Sci 16(4):323–343
Li X, Yeh AGO (2002b) Urban simulation using principal components analysis and cellular automata for land-use planning. Photogramm Eng Rem Sens 68(4):341–351
Li X, Yang QS, Liu XP (2007) Genetic algorithms for determining the parameters of cellular automata in urban simulation. Sci China SerD Earth Sci 50:1857–1866
Liao YL, Wang JF, Meng B, Li XH (2010) Integration of GP and GA for mapping population distribution. Int J Geogr Inf Sci 24(1):47–67
Liu Y (2008) Modelling urban development with geographical information systems and cellular automata. CRC Press, New York
Liu Y, Phinn SR (2003) Modelling urban development with cellular automata incorporating fuzzy-set approaches. Comput Environ Urban Syst 27(6):637–658
Lorena LAN, Furtado JC (2001) Constructive genetic algorithm for clustering problems. Evol Comput 9(3):309–327
Schmitt LM, Nehaniv CL, Fujii RH (1998) Linear analysis of genetic algorithms. Theor Comput Sci 200(1–2):101–134
Srinivas M, Patnaik LM (1994) Adaptive probabilities of crossover and mutation in genetic algorithms. IEEE Trans Syst Man Cybern 24(4):656–667
Stevens D, Dragićević S (2007) A GIS-based irregular cellular automata model of land-use change. Environ Plann B Plan Des 34(4):708–724
Stevens D, Dragićević S, Rothley K (2007) iCity: a GIS-CA modelling tool for urban planning and decision making. Environ Model Software 22(6):761–773
White RW, Engelen G (1993) Cellular automata and fractal urban form: a cellular modelling approach to the evolution of urban land use patterns. Environ Plann A 25(8):1175–1199
Wu F (1998) Simulating urban encroachment on rural land with fuzzy-logic-controlled cellular automata in a geographical information system. J Environ Manage 53(16):293–308
Wu F (2002) Calibration of stochastic cellular automata: the application to rural-urban land conversions. Int J Geogr Inf Sci 16(8):795–818

Chapter 4
Exploratory Spatial Data Analysis of Regional Economic Disparities in Beijing During 2001–2007

Xiaoyi Ma and Tao Pei

Abstract The unbalanced development of regional economics in Beijing has induced ubiquitous regional economic disparities which may affect the sustainable development of the economy and social stability. How to alleviate this unbalanced status is an important issue faced by both society and government. As would be expected the impact of the Olympic Games had the potential to boost the economy of the host city, however it is not clear whether such an event could narrow any regional economic disparities. In order to investigate the influence of the 2008 Beijing Olympic Games on the development of regional economics, this study examines the space-time dynamics of the economic development of Beijing from the year in which Beijing were successful in their bid (2001) using exploratory spatial data analysis (ESDA). The analysis was based on the county level and measure index of the regional per capita gross domestic product (GDP). The results do not show strong evidence of global spatial autocorrelation, but do give clear evidence of local spatial autocorrelation and spatial heterogeneity in the distribution of regional per capita GDP. The total economic disparity in Beijing was not alleviated by the drive to win the bid of the 2008 Olympics, and the financial and social situations became even more complicated. The speed of economic increase of the Changping and Shijingshan Districts was significantly lower than some neighboring regions, causing a new central polarization scheme gradually replaced the North–south polarization scheme that existed in Beijing between 2001 and 2007.

X.Y. Ma (✉) • T. Pei
State Key Laboratory of Resources and Environmental Information System, Institute of Geographical Sciences and Natural Resources Research, CAS, 11A Datun Road Anwai, 100101 Beijing, P.R. China
e-mail: maxy@lreis.ac.cn; peit@lreis.ac.cn

A. Yeh et al. (eds.), *Advances in Spatial Data Handling and GIS*,
Lecture Notes in Geoinformation and Cartography,
DOI 10.1007/978-3-642-25926-5_4,

4.1 Introduction

The unique historical foundation and location conditions caused an unbalanced regional economic development in Beijing. This unbalanced status induced ubiquitous regional economic disparities which had the potential to affect the sustainable development of economy and social stability. How to alleviate this regional status has become an important issue for both society and the government. History has indicated that to varying degrees, holding the Olympics can boost the economy of the host cities. The 2008 Olympic Games provided Beijing with an opportunity to successfully promote the economic development and accelerate the completion of major infrastructure upgrades in public service industries, transportation and other sectors. According to the research report by the Beijing Municipal Statistic Bureau, the total investment for the 2008 Olympic Games from 2002 to 2007 was about 351 hundred million dollars. This sum efficiently promoted the GDP growth of Beijing. In 2006, Beijing's per capita gross regional product reached 6,210 dollars exceeding the target of 6,000 dollars at the end of 2008, 2 years ahead of previously predicted. However, whether or not the Olympic Games could lessen the regional economic disparities was unclear. In order to investigate the influence of the 2008 Olympic Games on the development of regional economic disparities in Beijing, a study, reported in this paper, on the space-time dynamics of the economic development of Beijing after winning the Olympic bid, 2001 has been made. Exploratory spatial data analysis (ESDA) has been used.

There are a number of methods for identifying the spatio-temporal disparities between geographic objects, such as the Knox Test (Knox 1964), Jacquez k-Nearest Neighbour (Jacquez 1996), ST-DBSCAN (Wang et al. 2006), ESDA (Anselin 1994) and WNN (Pei et al. 2010). Among these methods, ESDA is the one which can emphasize the significance of spatial interactions and the geographical location. As a result, in the studies of regional economic development, we can use ESDA to characterize the economic performance of a region via identifying the spatial autocorrelation and spatial heterogeneity. Therefore, ESDA is a powerful tool for revealing the development of regional economic disparities. Several previous studies have been implemented on this issue focusing on the EU regions. Le Gallo and Ertur (2003), López-Bazo et al. (2004) applied ESDA to study the distribution of regional per capita GDP in Europe. Ertur and Koch (2006) investigated the case affected by the enlargement of the European Union between 1995 and 2000. In addition, some studies have assessed the regional economic disparities in China, using ESDA, including the work of Pu et al. (2005) on the Jiangsu province and Qiu et al. (2009) on the Huaihai economic zone of China. However, few revealed the space-time dynamics of regional economics caused by important historical events or national policy guides, particularly those of China.

This study, combining ESDA with GIS technology, investigates the development of regional economics after 2001, and attempts to explore the possibility that important historical events or national policy guides may have an association with change in spatial patterns of regional economic disparities over time. The method used in this study is based on the county level and measure index of the regional per capita gross domestic product (GDP).

4.2 Methodology and Data

4.2.1 Exploratory Spatial Data Analysis (ESDA)

Exploratory spatial data analysis (ESDA) is a set of techniques to describe and visualize spatial distributions, discover patterns of spatial association (spatial clustering or hot spots), identify atypical observations (outliers), and suggest different spatial regimes or other forms of spatial heterogeneity (Anselin 1994, 1996, 1999). Central to ESDA is the measure of global and local spatial autocorrelation.

The measure of global spatial autocorrelation is usually based on Moran's **I** statistic, expressed as:

$$I = \frac{n}{\sum_{i=1}^{n}(x_i - \bar{x})^2} \frac{\sum_{i=1}^{n}\sum_{j=1}^{n} w_{ij}(x_i - \bar{x})(x_j - \bar{x})}{\sum_{i=1}^{n}\sum_{j=1}^{n} w_{ij}} \tag{4.1}$$

where $\boldsymbol{n}$ is the number of regions in the study area, $\boldsymbol{x_i}$ and $\boldsymbol{x_j}$ are the attribute values for region $\boldsymbol{i}$ and $\boldsymbol{j}$, $\bar{x}$ is the mean attribute value for the $\boldsymbol{n}$ regions, and $\boldsymbol{w_{ij}}$ is the spatial weight matrix, which is usually defined as a bivariate symmetric matrix $\boldsymbol{W}$ to express the spatial contiguity relationship of regions. This paper adopts the spatial weight matrix based on the distance between the centroids of regions, that is, the corresponding element $\boldsymbol{w_{ij}} = 1$ if the distance between regions $\boldsymbol{i}$ and $\boldsymbol{j}$ is within the threshold distance and zero, otherwise.

The Moran's $\boldsymbol{I}$ statistic is based on the measure of covariance $(x_i - \bar{x})(x_j - \bar{x})$, which is used to assess the similarity of specified attribute values of variable $\boldsymbol{X}$ (Getis and Ord 1996). The inference of significance can be based on the permutation approach (Anselin 1995).

Moran's I statistic can be visualized as the slope in the Moran scatter plot (Anselin 1995), which plots the spatially lagged variable (***Wy***) on the original variable (***y***), here the ***y*** is in deviation from their mean. The four different quadrants of the Moran scatter plot represent the four types of local spatial association. The upper right (HH) and the lower left (LL) quadrants represent positive spatial autocorrelation, that is, a region with a high (low) value surrounded by regions also with high (low) values. The upper left (LH) and lower right (HL) quadrants correspond to negative spatial autocorrelation, i.e., a region is surrounded by regions with dissimilar values.

As one of the local spatial autocorrelation tools, the Moran scatter plot can identify local spatial clustering with high or low values, and indicate what patterns of spatial association make more contribution for the results of global spatial autocorrelation statistic, and visualize atypical observations. It makes up the major shortcoming of global spatial autocorrelation statistic, which can only indicate overall clustering. However, the Moran scatter plot does not indicate significance.

Local indicators of spatial autocorrelation (LISA) defined by Anselin (1995) provides a way to assess significant local spatial patterns, besides it can be used in a similar way to the Moran scatter plot.

The local Moran's statistics is a familiar LISA formulated as follows:

$$I_i = \frac{(x_i - \bar{x})}{m_0} \sum_j w_{ij}(x_i - \bar{x}) \; with \; m_0 = \sum_i \frac{(x_i - \bar{x})^2}{n} \tag{4.2}$$

where $\boldsymbol{n}$ is the number of regions in the study area, $\boldsymbol{x_i}$ is the attribute values for region $\boldsymbol{i}$, $\bar{x}$ is the mean of the attribute values for the $\boldsymbol{n}$ regions, and $\boldsymbol{w_{ij}}$ is the spatial weight matrix. The inference of significance is based on the permutation approach (Anselin 1995). The results of the local Moran's statistics can be classified into four spatial association types and visualized in a LISA cluster map (Anselin 1995).

4.2.2 Data and Background

The data for exploratory spatial data analysis has two types:

1. Regional per capita GDP and vital statistics data over the period 2001–2007 extracted from the Beijing area statistical yearbooks.
2. The administrative map of Beijing obtained from the administrative map of China with the scale of 1:1,000, 000 as shown in Fig. 4.1.

The venues of the Beijing Olympics were distributed into four main areas, one central area and three partitional areas. The central area is located in the Olympic park in the Chaoyang district, and the three partitions are the university area, the western community, the northern scenic tourist area, located in Haidian, Shijingshan, and Shunyi respectively. Figure 4.1 depicts the administrative map of Beijing with the gray area as venue regions and the light gray areas as the regions adjacent to the venue regions, according to the spatial weight matrix.

The theory is that venue and near-venue regions would be positively impacted by the presence of the Olympic Games. The per capita GDP growth of near-venue regions might be greater than those regions not close to Olympic events. Thus, some changes in spatial patterns of regional economic disparities could be expected to be observed due to the influence of the Olympic Games. Therefore, in the next section, ESDA is adopted to investigate the influence of the Olympic Games for the development of regional economic, and reveal the spatial pattern of regional disparities in Beijing.

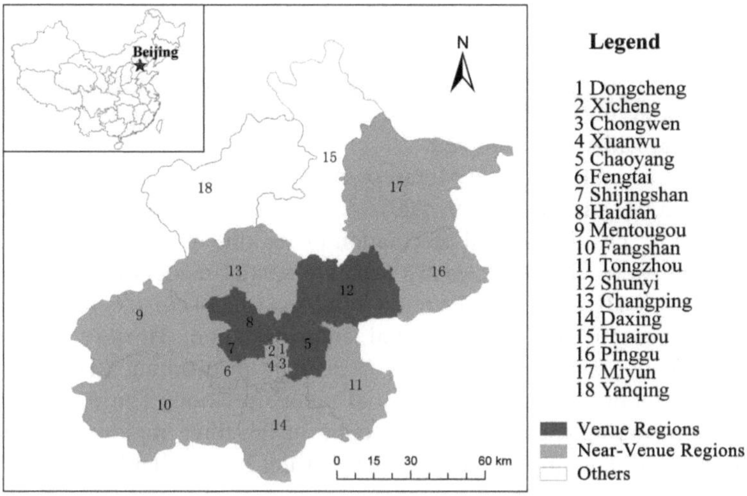

Fig. 4.1 The administrative map of Beijing, China

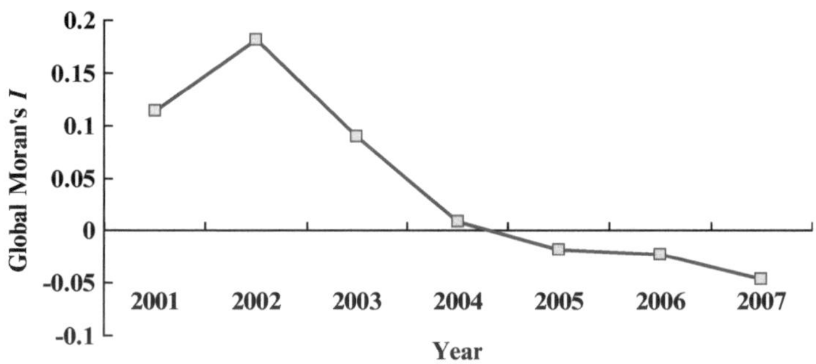

Fig. 4.2 The chart of Moran's I statistics for per capita GDP of Beijing's 18 districts or county, 2001–2007

4.3 Results and Analysis

In this study, both global and local spatial autocorrelation analyses were carried out by using the GeoDa software package (Anselin et al. 2006). Figure 4.2 shows the global Moran's ***I*** statistic for per capita GDP from 2001 to 2007 using the spatial weight matrix with the threshold distance about 36,000 m. The values of global Moran's ***I*** statistic show a downward trend between 2001 and 2007 and no significant positive or negative spatial autocorrelation, except for that in 2002. The results suggest that the locations of regions with high or low per capita GDP appear to be almost random, with a spatial pattern dominated by a developing trend of negative spatial autocorrelation from 2005.

The Moran's ***I*** is a global statistic, and gives a single result to assess the spatial association pattern for the entire study area. Therefore, it cannot identify the regions contributing more or less to the results of global spatial autocorrelation, and detect the hot spot or atypical localizations. In order to solve these problems, the Moran scatter plot and LISA were adopted.

Figures 4.3 and 4.4 show the Moran scatter plots of the per capita GDP for the initial year 2001 and final year 2007. It can be seen that 11 regions were characterized by positive spatial association (7 regions in quadrant HH, 4 in LL) and 7 regions were characterized by negative spatial associations (6 regions in quadrant LH, 1 in HL) in 2001. The regions of local spatial clusters with high values contributed more to the global spatial autocorrelation. However, 10 regions presented negative spatial associations (8 regions in quadrant LH, 2 in HL) in 2007, which dominated the global spatial autocorrelation. The main change of spatial pattern during the 2 years was found in the three regions of interest. In 2001, Xicheng, Changping and Shijingshan districts had similar high values for per capita GDP to those of their neighbours. These areas were in quadrant HH. However in 2007, Xicheng, located in quadrant HL reached a higher level of economic development than that of their neighbours'. Contrarily, Changping and Shijingshan moved to the quadrant LH.

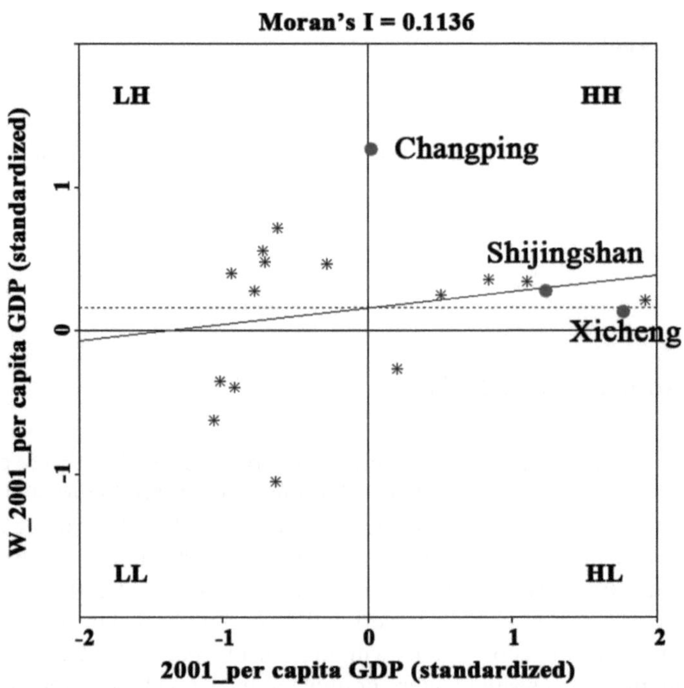

Fig. 4.3 Moran scatter plot for per capita GDP, 2001

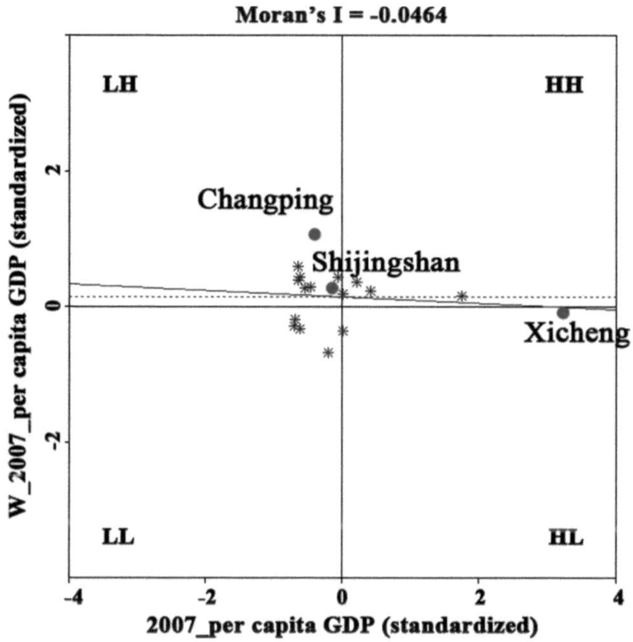

Fig. 4.4 Moran scatter plot for per capita GDP, 2007

The analysis of the Moran scatterplots appears to be contradictory to the theory that venue and near-venue regions would be positively impacted by the presence of the Olympic Games, and the per capita GDP growth of near-venue regions might be greater than those regions not close to Olympic events. Shijingshan District, as one of the venue regions, should have positive impact of the 2008 Olympics on the economic development. However, it was observed in Fig.4.4 that it was relatively poorer than surrounding regions by the year, 2007. The other venues and near-venue regions do not appear to have any obvious spatial pattern change.

Figure 4.5 shows the LISA cluster map for per capita GDP from 2001 to 2007. The coloured regions are significant with $p = 0.05$ for each year. The map indicates the two main patterns of spatial association, that is, clusters of high values are surrounded by high values (HH) and clusters of low values are surrounded by high values (LH). Since 2003, Changping and Shijingshan had moved one after the other from HH to LH region, which contributed to the creation of a new centre-surrounded polarization scheme that gradually replaced the original North–south polarization scheme.

The following interesting result is worth stressing: Although Shijingshan is one of the venue regions, and the Changping district is adjacent to both Haidian and Chaoyang (two venue regions), they belong significant to the LH quadrant and lagged behind the surrounded regions. This result could be explained that the bid winning of 2008 Olympics accelerated the proceedings of Suburbanization of Residence in Beijing. The large bedroom communities such as Tiantongyuan and

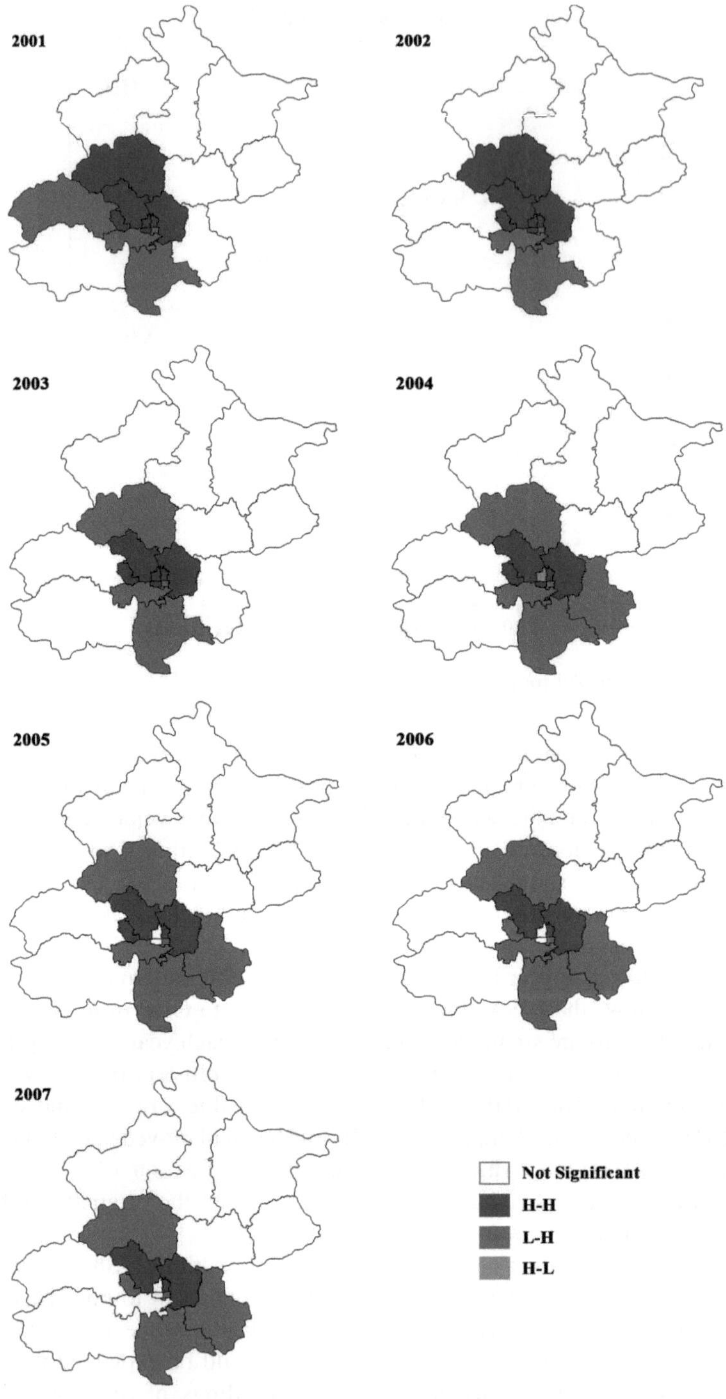

Fig. 4.5 The LISA cluster map for per capita GDP of Beijing's

Huilongguan were built in Changping district, which led to the sharp increase of population migration from the city centre to Changping district. Although the 2008 Olympic Games promoted the growth in GDP of Changping district, its per capita GDP is much lower than neighbours' due to the increase of population. In addition, Shijingshan district also moved from the quadrant HH to the quadrant LH following Changping in 2006. An efficient cause should be the policy-based limiting production of the Capital Iron and Steel Company. Due to the 2008 Olympic Games, Beijing accelerated the improvement of environment pollution by decreasing the production of the Capital Iron and Steel Company (Wan 2006), which led to the lower GDP increasing in Shijingshan than some surrounding areas.

4.4 Conclusions

This study has investigated the influence of the 2008 Olympic Games as regards the development of regional economic disparities in the city, from when Beijing won their bid to host the Olympic Games (2001). By combining ESDA with GIS technology, the spatial association patterns of regional per capita GDP were analyzed between and including 2001 and 2007.

The results showed weak evidence of global spatial autocorrelation, but clear evidence of local spatial autocorrelation and spatial heterogeneity in the distribution of regional per capita GDP. Although the 2008 Olympic Games boosted the economic development of Beijing, the total economic disparity in the various areas of Beijing was not alleviated.

The results of the Moran scatterplot and the LISA highlight that in Beijing from 2001 to 2007, a new centre-surrounding polarization scheme gradually began replacing the once familiar North–south polarization.

Acknowledgments This study was funded through support from a grant (Project Number: 2006AA120106) from the High-Tech Research and Development Programme of China, a grant from the Chinese Academy of Sciences (KZCX2-YW-QN303), a grant from the Institute of Geographical Sciences and Natural Resources Research (Project Number: 200905004).

References

Anselin L (1994) Exploratory spatial data analysis and geographic information systems. In: Painho M (ed) New tools for spatial analysis. Eurostat, Luxembourg, pp 45–54

Anselin L (1995) Local indicators of spatial association — LISA. Geogr Anal 27:93–115

Anselin L (1996) The Moran scatterplot as an ESDA tool to assess local instability in spatial association. In: Fischer MM, Scholten HJ, Unwin D (eds) Spatial analytical perspectives on GIS. Taylor & Francis, London, pp 111–125

Anselin L (1999) Interactive techniques and exploratory spatial data analysis. In: Longley P et al (eds) Geographical information systems: principles, techniques, management and applications. Wiley, New York, pp 251–264

Anselin L, Syabri I, Kho Y (2006) GeoDa: an introduction to spatial data analysis. Geogr Anal 38:5–22

Ertur C, Koch W (2006) Regional disparities in the European Union and the enlargement process: an exploratory spatial data analysis, 1995–2000. Ann Reg Sci 40:723–765

Getis A, Ord J (1996) Local spatial statistics: an overview. In: Longley PA, Batty M (eds) Spatial analysis: modeling in a GIS environment. Geoinformation International, Cambridge, pp 261–277

Jacquez GM (1996) A k nearest neighbour test for space–time interaction. Stat Med 15:1935–1949

Knox EG (1964) The detection of space–time interactions. Appl Stat 13:25–30

Le Gallo J, Ertur C (2003) Exploratory spatial data analysis of the distribution of regional per capita GDP in Europe, 1980–1995. Reg Sci 82:175–201

López-Bazo E, Vayá E, Artís M (2004) Regional externalities and growth: evidence from European regions. Reg Sci 44:43–73

Pei T, Zhou CH, Zhu AX, Li BL, Qin CZ (2010) Windowed nearest-neighbour method for mining spatio-temporal clusters in the presence of noise. Int J Geogr Inf Sci 24(6):925–948

Pu Y, Ge Y, Ma R et al (2005) Analyzing regional economic disparities based on ESDA. Geogr Res 24(6):965–974

Qiu F, Zhu C, Tong L et al (2009) Spatial analysis of economic disparities of county level in Huaihai Economic Zone. Sci Geogr Sin 29(1):56–63

Wan X (2006) Report on the implement of national economic and social development plan in 2006 and draft in 2007 of Shijingshan district. http://huigu.bjsjs.gov.cn/affair/guihuajh/8a8481d21b533410011b584631130012.html

Wang M, Wang AP, Li AB (2006) Mining spatial-temporal clusters from Geo-databases. Lect Notes Artif Int 4093:263–270

Chapter 5
An Approach to Discovering Spatial–Temporal Patterns in Geographical Processes

Siyue Chai, Fenzhen Su, and Weiling Ma

Abstract Spatial data mining focuses on searching rules of the geographical statement, the structures of distribution and the spatial patterns of phenomena. However, many methods ignore the temporal information, thus, limited results describing the statement of spatial phenomena. This paper focuses on developing a mining method which directly detects spatial–temporal association rules hidden in the geographical process. Through such approach, geographical process can be extracted as a particle which exists in spatial–temporal-attribute dimensions. By setting customized fixed-window, geographical process in one time interval is organized as a record with attribute value and spatial orientation change. Spatial–temporal association rules can be found in geographic process mining table.

$$[\text{TimeInterval}_i, \text{MovingDirection}_m, \text{P}] \Rightarrow [\text{TimeInterval}_j, \text{MovingDirection}_n, \text{Q}]$$

To verify this mining approach, it is applied on AVHRR MCSST thermal data for extracting Indo-Pacific warm pool's frequent movement patterns. The raw data provided by PO.DAAC, whose time spans of 20 years from 1981 to 2000 with 7 days' time particle, has been used to mining spatial temporal association rules. In the experiment, we extract warm pool within 30°N–30°S, 100°E–140°W and use 28°C as temperature threshold. After which Warm Pool's geographical process table is established so as to describe the variation of warm pool in spatial–temporal-attribute dimension. In the mining process, 18 spatial–temporal process frequent models can be found by setting minimal support threshold at 10% and confidence

S.Y. Chai • W.L. Ma
Institute of Geographic Sciences and Natural Resources Research, Beijing, 100101, China

Chinese Academy of Science, Beijing 100101, China

F.Z. Su (✉)
Institute of Geographic Sciences and Natural Resources Research, Beijing, 100101, China
e-mail: sufz@lreis.ac.cn

A. Yeh et al. (eds.), *Advances in Spatial Data Handling and GIS*,
Lecture Notes in Geoinformation and Cartography,
DOI 10.1007/978-3-642-25926-5_5, © Springer-Verlag Berlin Heidelberg 2012

threshold at 60%. The result shows such a methodology can mine complicated spatial–temporal rules in realistic data. At the same time, the mining result of warm pool's frequent movement patterns may provide reference for oceanographers.

5.1 Introduction

Association Rule Mining (ARM) is a popular and prevailing research method for discovering interesting relationships between variables in large databases. Previously, Spatial Association Rules Mining (SARM) and Sequence Mining (SM) were two different subareas dealing with spatial rules and temporal rules respectively. SARM is concerned with connection rules which represent spatial topology, such as "is_Close" and "is_Next_To", among spatial objects (Koperski and Han 1995). So a spatial association rule (SAR) is an association rule where at least one of the predicates is spatial (Shekhar and Huang 2001). SM is concerned with finding association rules under temporal constraints (Agrawal and Srikant 1995; Zaki 2001). However, with the development of information capture devices, data recorded contain both time stamp and location. Especially, when Process Geographical Information System (PGIS) (Fen-zhen and Cheng-hu 2006) is under discussion, the mining of spatial–temporal rules can be expected. Therefore a burning question is how to find a solution that can mine spatial–temporal association rules.

From a research objectives prospective, spatial–temporal association rules (STARs) can be divided into two types. One is to find uniformity rules among group objects, the other focuses on mining movement patterns of one object's flexible shape, such as the 'warm pool'. In this paper, we propose an approach that takes into account spatial–temporal attributes when miming frequent trajectory patterns of flexible shaped objects. The approach contains three steps. Fundamentally, an entire database is scanned and a geographic process organized. Then the pre-processing is required for process recording. Finally, the a-priori algorithm (Agrawal et al. 1993) is applied to mine spatial–temporal association rules. To validate such an approach, an experiment was established using an AVHRR remote sensing image provided by PO. DAAC, whose time spans is 20 years from 1981 to 2000 with a 7 days time capsule, to mine the most frequent trajectories of the Indo-Pacific 'warm pool'. Some interesting rules have been found which comply with the general rules of the Warm Pool seasonal fluctuation, while leaving some rules still unresolved.

5.2 Related Work

From an analytical perspective, the target of STARs mining is to search for the correlations hidden behind object spatial and temporal data. Automatic statistic approaches are used for extracting such correlations from two-dimensional table structure which covers realistic object statements or abstracted object progress.

Such a statistical approach is well developed using related database techniques and traditional data mining algorithms. In traditional data mining, R. Agrawal et al. (Agrawal et al. 1993) proposed an a-priori algorithm to mine frequent item sets from a transactional database. The form of the rule is abstracted as X=>Y. The algorithm employs prior knowledge of frequent item set properties, which are all non-empty subsets of a frequent item set that must also be frequent. Under such prior knowledge, the number of candidate set combinations is produced prominently. A-priori algorithm established a milestone in ARM. Two years later, J. Han et al. (Han and Fu 1995) developed Agrawal's research and proposed an algorithm which utilized a multi-level logic structure to abstract target rules and mined cross level schema so that complicated logic rules could be described.

After such fundamental work, many data mining researchers (Agrawal and Srikant 1995; Pei et al. 2001; Zaki 2001; Hwang et al. 2004; Winarko and Roddick 2007; Kong et al. 2010) demonstrated how to combine both temporal information and other attributes into sequence rules and temporal rules. In rich temporal description forms, from basic time stamp representation to process representation, temporal topology was introduced. To take account of data structure, Zaki's SPADE algorithm decomposed the original mining problem into several smaller sub-problems so that each sub-problem could be independently solved in memory, while J. Pei et al.'s PrefixSpan algorithm reduced the search space and accelerated the mining process by using projected data table. In temporal topology, Kong et al. proposed an algorithm which used four kinds of temporal topology predicates, namely before, during, equal to and overlapping, to mine temporal association rules. Despite temporal topology, Winarko et al. proposed a method called ARMADA, which accommodating temporal intervals offered rules that were richer still.

Geographic researchers discussed ways of taking advantage of multi-level logic structures and cross mining schema in Geographic Information System so as to automatically find dependent patterns in spatial data. Han et al. (Koperski and Han 1995) set multilevel concept descriptions for spatial rules, such as "not_disjoint, close_to", "Intersects, Inside, Contain, Equal" to show the spatial topology. The final rules can be represented as "is_a(x, house) A close_to(x, beach) → is_expensive(x) (90%)". Their work made spatial information and none spatial–temporal information consistent. The AGM algorithm (Inokuchi et al. 2000) and the FSG algorithm (Kuramochi and Karypis 2001) used a-priori approach to combine frequent sub-graphs mined at the previous level to generate all candidates at the next level. Appice et al. (Appice and Buono 2005) summarizes the advantages gained from this taxonomic knowledge on spatial data, to mine multi-level spatial association rules. Bembenik et al. (Bembenik and Rybiński 2009) defined the neighborhood in terms of Delaunay diagrams, instead of predefining distance thresholds with extra runs.

However, how to handle both spatial and temporal data is still a subject of research. Lee and Chan (Lee et al. 2009) developed an approach using a TI-lists structure and a GBM algorithm to find trajectory patterns. Huang et al. (Huang et al. 2008) summarized the correlation between sea salt and temperature in spatial–temporal distribution. The final rules can be described as "if the salinity rises from 0.15 to 0.25 psu in an area lying in an east–northeasterly direction and is near Taiwan, then the temperature will rise from 0°C to 1.2°C in the area that is in a east–northeast direction and is far away from Taiwan the following month", which set temporal constraints that limit temporal information in two adjacent time intervals. Verhein et al. (Verhein and Chawla 2008) proposed a source-thoroughfare-sink model describing trajectory patterns, while leaving charts without connections unexplained. In conclusion, traditional algorithms may lose important information during the search for frequent trajectory patterns exhibited by flexible shaped objects.

5.3 Geographic Process Mining Approach

This section describes the concepts associated with association rule mining to discover spatial–temporal patterns. Some definitions are given before describing the proposed algorithm. Figure 5.1 below shows the processing steps.

5.3.1 Description of Geographic Process

Modern physics shows that time and space form an indivisible entirety called four-dimension space. In this space, continuity dominates the attributes of existing objects (at low speeds), which means object snap shots always exists. In geography, modern science provides powerful devices to capture snapshots of geographic phenomena rapidly and continuously. Therefore, the mining of geographic patterns from these primary data needs strict organization and definitions of spatial–temporal statements.

Definition 5.1. In four-dimensional space, given an exact T_i, objects *I* can be represented as $I = [\mathrm{Time}_i, \mathrm{Location}_i, \mathrm{Attribute}_i]$, $i \in N$, $T_i \in (-\infty, -\infty)$. Location is an object spatial vector, usually Location = [Longitude, Latitude, Altitude]^T, Attribute is an object's attribute vector.

Every snapshot can be described as $I = [T_i, L_i, A_i]$. Attribute A is a many dimensions vector, depicting an object from different aspects. To represent the geometry of a flexible appearance object, indices can be used, such as degree of fragmentation, fractal measures etc. (Coster and Chermant 1985). Therefore, flexible shaped objects can be recorded via appropriate kinds of index. In addition a spatial–temporal description statement is compatible with a two-dimensional table structure, which means the characterization of an object can be stored within a formal database table (Table 5.1).

Definition 5.2. A Geographical Process (GP) is the change from a starting statement to an ending statement. Given a fixed time window TI, $TI = [T_i, T_j]$, $i < j$ and $i, j \in N$, $GP = {}^{\Delta}I = [TI, {}^{\Delta}L, {}^{\Delta}A]$.

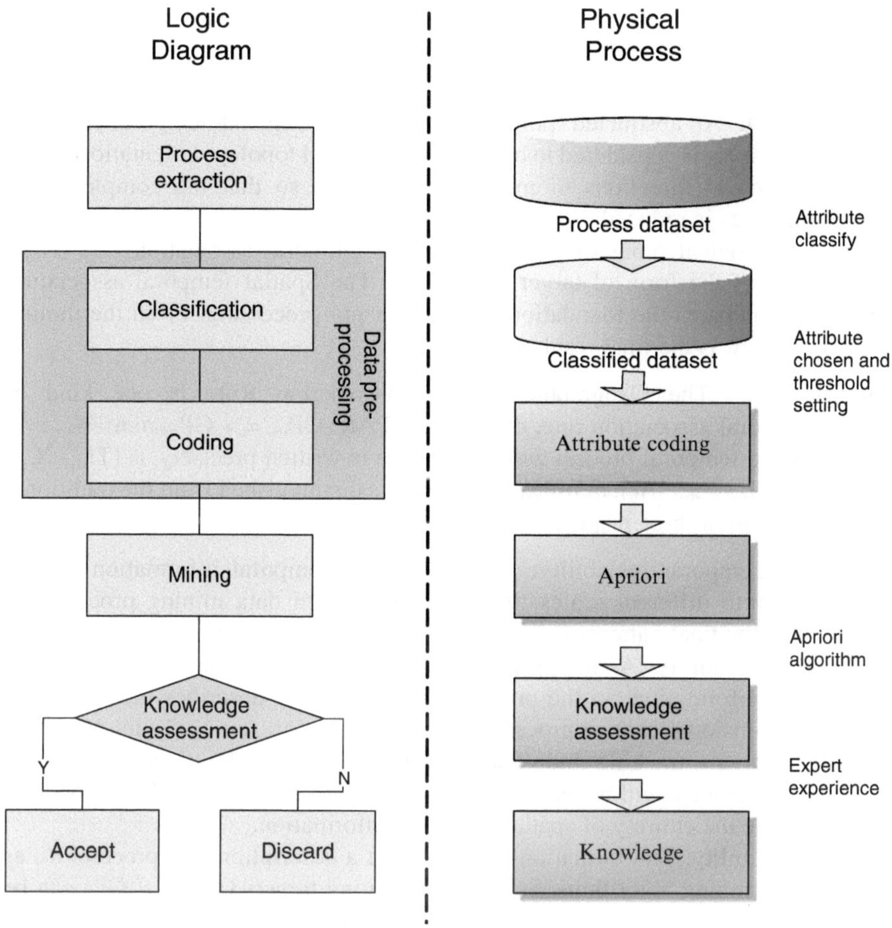

Fig. 5.1 Mining approach flow chart

Table 5.1 An example of spatial–temporal statement table

ID	Time	Longitude	Latitude	Temperature	Area (10^6 km^2)
1	1981-11-3	156.62	2.61	28.96	42.55
2	1981-11-10	158.57	1.43	29.44	42.1
3	1981-11-17	160.9	0.17	29.47	42.42
4	1981-11-24	161.46	−1.17	29.23	40.77
5	1981-12-1	163.04	−1.50	29.33	38.84

Definition 5.3. The Time Interval in a GP relates to some 'particle', PP, undergoing a process.

Geographic process, which reflects an object's spatial movement as time evolves, is composed of sets of quantification and qualification statements.

An Object's spatial change $^{\Delta}L$ data set contains change of shape information and moving location information. To simplify such variety, the target object is viewed as a point.

Definition 5.4. An abstracted 'point' object is called a spatial–temporal point.

Since an object is considered to be a point, the spatial topology orientation model is limited to just directions of moving (Table 5.2) so that the complexity of association rules is reduced.

From the logical point of view, the spatial–temporal association rule is the instantiation of the form of target knowledge. The Spatial–temporal association rule format prepares the foundation data set for pre-processing. From the mining point of view, it is the final result goal.

Definition 5.5. The Geographic Process Association Rule is one kind of spatial–temporal association rule, defined as GPAR: $GP_m \Rightarrow GP_n$, $m,n \in N$.

The spatial temporal process association rule is written precisely as $[TI_m, {}^{\Delta}L_m, {}^{\Delta}A_m] = >[TI_n, {}^{\Delta}L_n, {}^{\Delta}A_n]$, $m,n \in N$. Such a rule is distinguished from the traditional association rule in three respects.

(a) Spatial–Temporal scalability: The spatial and temporal information can be divided into different scales. In a question-driven data mining process, the detail of the final patterns is influenced by the spatial–temporal particle.
(b) Evolution entirety: As the process association rule contains spatial–temporal coupled information, so the process association rule describes the integrated four-dimensional change process. In this experiment, setting 3 months as the time intervals tracking Indo-Pacific warm pool particle movements every statement change will be recorded and each attribute will be combined together to provide the entirety of spatial–temporal information.
(c) Rule flexibility: The formation of rule is just a description of a process. Rules separate mining algorithms. So different mining targets $GP_m \Rightarrow GP_n$ can be interpreted as carrying diverse meanings. To take advantage of these characteristic, different mining algorithms shares the same processes as data sets.

Table 5.2 An example of spatial–temporal process table (PP = 3 months)

TIID	Direction	Temperature	Area (10^6 km^2)	Start time	End time
1	N	−0.10	−6.98	1981 Winter	1982 Spring
2	NE	−0.24	4.50	1982 Spring	1982 Summer
3	SE	0.09	1.25	1982 Summer	1982 Autumn
4	SW	−0.06	−7.72	1982 Autumn	1982 Winter
5	NW	−0.26	−2.04	1982 Winter	1983 Spring

5.3.2 Geographic Process Data Pre-processing

This section contains three parts. The pre-processing approach is discussed. Then, how to organize the Geographic Process Mining Table (GPMT) through pre-processing is mentioned. As a target GPAR rule (discussed above), a mining method has been developed based on a-priori algorithm.

5.3.2.1 Process Data Set Pre-processing

Data pre-processing is an approach to extract data from a primary concept level to upper levels. It is a question-driven procedure. All attributes of a spatial–temporal particle, which directly influences the final rule statistics, are classified. After extracting the dataset of a geographic process, one record in the table is called a meta-process. Different attributes have different kinds of classification threshold in relation to the raising of its concept level. Such as in Table 5.3, using a variance based binning method, to classify "Temperature". Three bins are established representing "fall", "remain practically unchanged", "rise" in turn. The result is shown below.

Reorganization and conversion makes every subset of the process consistent (Fig. 5.2). In the process employed in this experiment [TI, $^{\Delta}$L, $^{\Delta}$A], using an orientation model, the number of changing domain "$^{\Delta}$L" sets was 8, and the number of changing domains "A" were 3. So taking the first row in Table 5.2 as an example, the result saved is coded "421" (Table 5.4).

5.3.2.2 Building the GPMT

Definition 5.6. The GPAR's Time Span (TS) is the time span between TI_m and TI_n. So the TS∈[1,K], K = (Starting Time − Ending Time)/PP,K∈N.
The temporal connection hidden in the GPAR: $GP_m \Rightarrow GP_n$ is limited. For example, in this experiment, the maximum TS is 75 in 20 years. To mine such GPAR, GPMT is constructed by connecting the process dataset with a cloned process dataset with TS time interval lags (Fig. 5.3).

Table 5.3 Classification after binning

Class	Lower threshold	Upper threshold
0		<−0.1140654
1	≥−0.1140654	≤0.1126115
2	>0.1126115	<0.067004

Fig. 5.2 Coding format

Meta process: XXXXXXXXXX
└X is classification number

Table 5.4 Process dataset

TI_ID	Process code	Start time	End time
1	111	1981 Winter	1982 Spring
2	201	1982 Spring	1982 Summer
3	411	1982 Summer	1982 Autumn
4	610	1982 Autumn	1982 Winter

Fig. 5.3 Geographic process mining table (TS = 1 process)

5.3.3 *Process Association Rules Data Mining*

In this section, a-priori algorithm (Agrawal et al. 1993) introduced to mine GPARs based on GPMT. By setting Maximum length association rules and Support threshold = min_sup (support(A=>B) = P(A∩B)), confidence threshold = min_conf (confidence(A=>B) = P(B|A)), a-priori algorithm process dataset traverse time is not larger than the set maximum length times. After the first Scan, a-priori captures frequent set rule Length equals 1 and whose support and confidence are not less than min_sup and min_confidence. Then Length + 1 candidate, is built and then Scan DB is used for a second times, using a-priori to prune false candidates, and generate Length + 1 candidates. This procedure continues until no Length + 1 frequent sets can be found or Length equals Maximum length.

5.3.3.1 Pseudocodes

(1) L1 = {large 1-item sets} //Scan DB once, capture frequent set, L1
(2) for (k = 2;$L_{k-1} \neq \varphi$; k++)
(3) {C_k = apriori-gen($L_{k\ 1}$)//according to L_{k-1} frequent set, digging new candidate sets, Ck
(4) for each t ∈ D
(5) {Cspt = subset (C_k,t);
(6) for each c ∈ C_t c.count ++;
(7) }
(8) Lk = {c ∈ C_k|c.count ≥ minsupp}
(9) }
(10) L = $\cup_k L_k$;

5.4 Experimental Results and Analysis

In this section realistic remote sensing images are presented and the spatial temporal association mining rules generated from them. To analyze the results, some work by oceanographers is quoted to assess the mining rules.

5.4.1 SST Data Introduction

These experiments use AVHRR sensor SST to analyze the data provided by PO. DAAC whose size is 2048 × 1024 pixel^2. Under Mercator's projection, spatial resolution is 5.689 pixels per Longitude and 2.845 pixels per Latitude. Temporal resolution is 7 days and the data covers 20 years (1981–2001). According to Wang et al. (Wanying 1998), the Indo-Pacific Warm Pool (WP) envelope should be 100°E–140°W, 30°N–140°S within which the warm pool influences the weather mostly of Eastern Asia. Meanwhile, we use 28°C as the edge of the Warm Pool accepted by many oceanographers (Qilong and Xuechuan 1997; Enfield et al. 2005) to extract geometries (Fig. 5.4).

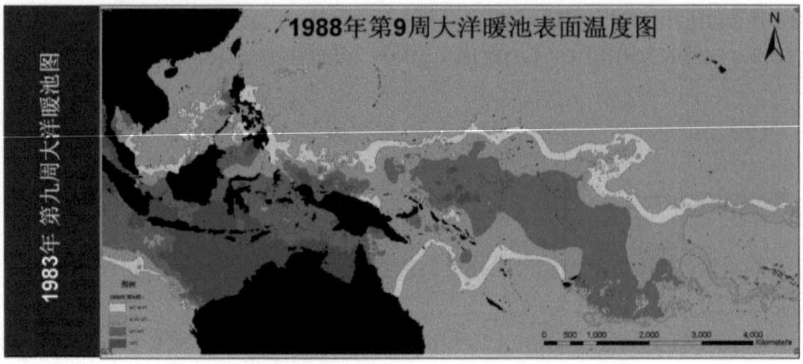

Fig. 5.4 Warm pool extraction image

Table 5.5 Seasonal process table: the top four GPARs results are listed

TI-ID	Direction	Temperature	Area change (10^6 km^2)	Start time	End time
1	1	−0.10	−0.70	Spring	Summer
2	2	−.24	4.50	Summer	Autumn
3	4	0.95	1.25	Autumn	Winter
4	6	−0.06	−7.72	Winter	Spring
5	7	−0.27	2.04	Spring	Summer

5.4.2 *Experiment*

Using the data processing approach mentioned above, the Warm Pool is described quantitatively in a Warm Pool spatial–temporal statement table which is to be captured by a sensor in 7 days. The maximum sequential rules number 3, the min_support is 10%, and confidence is 60%. The temporal scales are of 3 months so as to slice the data into large amounts.

5.4.2.1 Experiment

Four solar terms are chosen such as spring, summer, autumn and winter. So the splitting might be the Spring Equinox, Summer Solstice, Autumnal equinox and Winter Solstice. The seasonal processes recorded (Table 5.5) can be obtained from Table 5.2.

5.4.2.2 Rule Analysis

The El Niño-Southern Oscillation (ENSO), is a climate pattern that occurs across the tropical Pacific Ocean on average every 5 years, but a period which varies from 3 to 7 years, and is therefore, widely known as "quasi-periodic." The two components are coupled: when the warm oceanic phase (known as *El Niño*) is in effect, surface pressures in the western Pacific are high, and when the cold phase is in effect (*La Niña*), surface pressures in the western Pacific are low (Trenberth et al. 2007). The Southern Oscillation is an oscillation in air pressure between the tropical eastern and the western Pacific Ocean waters. Low atmospheric pressure tends to occur over warm water and high pressure over cold water, in part because deep convection over the warm water acts to transport air. Normal equatorial winds warm as they flow westward across the Pacific. Cold water is pulled up along the west coast of South America. Warming water is pushed toward the western side of the Pacific (http://en.wikipedia.org/).

The movement of warm water forms a periodical geographical process. After mining, GPARs may have association with the Southern Oscillation. Here is the comparison (Table 5.6).

GPAR1: If during autumn to winter, the Warm Pool moves south without prominent temperature and area changes, then in the next year's summer to autumn, Warm Pool had moved north without significant temperature and area change. This rule reflects the warm pool's annual latitude infatuation. Two none spatial–temporal attributes remain unchanged. The rules appeared in 1984, 1992, 1993, 1995 and 1997. Almost earlier than all these rules occurred the El Niño phenomena took place. El Niño happens between October and the following February. It occurred in 1982–1983,1986–1987, 1991–1992, 1993, 1994–1995, 1997–1998. These rules (events) occurred just after El Niño, excluding 1984. But during 1982–1983, one of the most powerful El Niño took place. So the connection between GPAR1 and El Niño has shown some kind linkage which needs further research and explanation.

GPAR2: If during spring to summer, the Warm Pool moves northwest with its Area shrinking significantly, then the following winter to spring, the Warm Pool will move south without significant temperature and area change. Unlike GPAR1, the rules applied in 1993, 1998, 1999, with 100% confidence and the warm pool dominant area shrank. In all dataset, during spring to summer, a shrinking Warm

Table 5.6 Mining results

Consequent	Antecedent	GPAR	Support %	Confidence %
Summer to autumn = 111	Autumn to winter = 511	[Summer to autumn, N, 1, 1] is_before [Autumn to winter, S, 1, 1]	27.8	80
Winter to spring = 511	Spring to summer = 810	[Winter to spring, S, 1, 1] is_after [Spring to summer, NW, 1, 0]	16.7	100
Autumn to winter = 512	Spring to summer = 810	[Autumn to winter, S, 1, 2] is_after [Spring to summer, NW, 1, 0]	16.7	66.7
Summer to autumn = 112	Spring to summer = 111	[Summer to autumn, N, 1, 2] is_after [Spring to summer, N, 1, 1]	16.7	66.7

Pool area just appeared in the 4 years, 1993, 1997, 1998, and 1999. This phenomena may be explained by too many rains over the sea or a powerful cold stream. It still needs oceangraphers to explain this.

GPAR3: In the same year, if during spring to summer, the Warm Pool moves northwest with its area shrinking significantly, then during autumn to winter then the Warm Pool will move south with area expanding . Such rules happened in 1998, 1999, with the pool dominant area shrinking in spring to summer, and expanding right into autumn to winter. In all datasets shrinking during spring to summer of the Warm Pool area just appeared in the 4 years, 1993, 1997, 1998, 1999, while during autumn to winter in 1988, 1989, 1996, 1998, and 1999, more rules would be generated if the data classification scale was expanded. However, this rule may also be just a coincidence because its support and confidence levels are not high enough.

GPAR4: In the same year, if during spring to summer, the Warm Pool moves north without significant temperature and area change then during summer and autumn, the Warm Pool will move north with area expanding significantly, just as for GPAR3. This rule needs more data for satisfactory proof.

Each of the GPARs listed here has high support and confidence. The index of lift which is defined as Confidence/Support, are all greater than 1, which means all these rules are strong rules where GP1=>GP2 are positively correlated. Despite these statistics, years when these GPARs occur may really correlate with El Niño and the Southern Oscillation. Many oceanographers have worked on this subject and found patterns connecting the Warm Pool and the Southern Oscillation (McPhaden and Picaut 1990; Brijker et al. 2007; Cheng et al. 2008).

5.5 Conclusions

In this paper, an approach is proposed for mining the spatial–temporal association rules of a moving flexible-shaped object. The approach consists of three steps: reorganizing data to form geographic process table; setting parameters to classify and code all attributes; finally, the spatial temporal association rules are mined using a-priori algorithm from coded GPMT. Although we have shown the rules mined by such an approach some issues need to be further addressed in future research. Above all, how to enrich spatial–temporal topologies so that more complicated rules can be represented way and can be dealt with is an advanced question. The GPARs described in this paper are just an example of spatial–temporal association rules. The results could also be used for other applications such as mobile advertisements, shopper trajectory analysis, and animal trajectory analysis to discover other interesting rules.

References

Agrawal R, Srikant R (1995) Mining sequential patterns. In: Proceedings of the 11th international conference on data engineering (ICDE'95), IEEE Computer Society, Taipei, pp 3–14

Agrawal R, Imielinski T, Swami A (1993) Mining association rules between sets of items in large databases. ACM SIGMOD Rec 22(2):207–216

Appice A, Buono P (2005) Analyzing multi-level spatial association rules through a graph-based visualization. In: Ali M, Esposito F (eds) Innovations in applied artificial intelligence, vol 3533. Springer, Heidelberg, pp 448–458

Bembenik R, Rybiński H (2009) FARICS: a method of mining spatial association rules and collocations using clustering and Delaunay diagrams. J Intell Info Syst 33(1):41–64

Brijker JM, Jung SJA et al (2007) ENSO related decadal scale climate variability from the Indo-Pacific warm pool. Earth Planet Sci Lett 253(1–2):67–82

Cheng X, Qi Y et al (2008) Trends of sea level variations in the Indo-Pacific warm pool. Global Planet Change 63(1):57–66

Coster M, Chermant JL (1985) Précis d'analyse d' images. Editions du CNRS, Paris

Enfield DB, Lee S-K et al (2005) How are large western hemisphere warm pools formed? Prog Oceanography 70(2–4):346–365

Fen-zhen SU, Cheng-hu ZHOU (2006) A framework for Process Geographical Information System. Geogr Res 25(3):477–483

Han J, Fu Y (1995) Discovery of multiple-level association rules from large databases. In: Dayal U, Gray PMD, Nishio S (eds) Proceedings of 21st international conference on very large data bases (VLDB'95), Morgan Kaufmann, Zurich, pp 420–431

Huang Y-P, Kao L-J et al (2008) Efficient mining of salinity and temperature association rules from ARGO data. Expert Syst Appl 35(1–2):59–68

Hwang S-Y, Wei C-P et al (2004) Discovery of temporal patterns from process instances. Comput Ind 53(3):345–364

Inokuchi A, Washio T et al. (2000) An apriori algorithm for mining frequent substructures from graph data. Principles of Data Mining and Knowledge Discovery 73(2):13–28

Kong X, Wei Q et al. (2010) An approach to discovering multi-temporal patterns and its application to financial databases. Information Sciences 180(6):873–885

Koperski K, Han J (1995) Discovery of spatial association rules in geographic information databases. In: Proceedings of the 4th international symposium on advances in spatial databases (SSD'95), Springer, Berlin, pp 47–66

Kuramochi M, Karypis G (2001) Frequent subgraph discovery. In: Proceedings of the IEEE international conference on data mining (ICDM'01), San Jose, pp 313–320

Lee AJT, Chen Y-A et al (2009) Mining frequent trajectory patterns in spatial–temporal databases. Info Sci 179(13):2218–2231

McPhaden MJ, Picaut J (1990) El Niñ-Southern Oscillation displacements of the western equatorial Pacific warm pool. Science 250:1385–1388

Pei J, Han J, Mortazavi-Asl B, Chen Q, Dayal U, Hsu MC (2001) PrefixSpan: mining sequential patterns efficiently by prefix-projected pattern growth. In: Proceedings of the 2001 international conference on extending database technology: advances in database technology, Heidelberg, pp 215–224

Qilong Z, Xuechuan W (1997) Analysis of some oceangraphic characteristics of the tropical western pacific warm pool. Studia Marina Sin 38(01):31–38. El Niño-Southern Oscillation, http://en.wikipedia.org/

Shekhar S, Huang Y (2001) Discovering spatial co-location patterns: a summary of results. In: Jensen CS, Schneider M, Seeger VJ, Tsotras B (eds) Proceedings of the 7th international symposium on the advances in spatial and temporal databases. LNCS, vol 2121. Springer, Berlin/Heidelberg/New York, pp 236–256

Trenberth KE, Jones PD et al (2007) Observations: surface and atmospheric climate change. In: Qin SD, Manning M et al (eds) Climate change 2007: the physical science basis. Contribution

of working group I to the fourth assessment report of the intergovernmental panel on climate change. Cambridge University Press, Cambridge, UK, pp 235–336
Verhein F, Chawla S (2008) Mining spatio-temporal patterns in object mobility databases. Data Min Knowl Discov 16(1):5–38
Wanying LKZCS (1998) Basic features of the warm pool in the western pacific and its impact on climate. Acta Geogr Sin 53(6):511–519
Winarko E, Roddick JF (2007) ARMADA – an algorithm for discovering richer relative temporal association rules from interval-based data. Data Knowl Eng 63(1):76–90
Zaki MJ (2001) SPADE: an efficient algorithm for mining frequent sequences. Mach Learn 42 (1):31–60

Part III
Generalization and Indexing

Chapter 6
Normalizing Spatial Information to Improve Geographical Information Indexing and Retrieval in Digital Libraries

Damien Palacio, Christian Sallaberry, and Mauro Gaio

Abstract Our contribution is dedicated to geographic information contained in unstructured textual documents. The main focus of this article is to propose a general indexing strategy that is dedicated to spatial information, but which could be applied to temporal and thematic information as well. More specifically, we have developed a process flow that indexes the spatial information contained in textual documents. This process flow interprets spatial information and computes corresponding accurate footprints. Our goal is to normalize such heterogeneous grained and scaled spatial information (points, polylines, polygons). This normalization is carried out at the index level by grouping spatial information together within spatial areas and by using statistics to compute frequencies for such areas and weights for the retrieved documents.

6.1 Introduction

The digitization of printed literature is currently making significant progress. The Google Books Library Project, for instance, aims at creating digital representations of the entire printed inventory of libraries. Other initiatives specialize in the legacy literature of specific domains, such as medicine or cultural heritage (Sautter et al. 2007). For instance, libraries or museums are now offering their electronic contents to a growing number of users.

While some projects only aim at creating digital versions of the text documents, domain-specific efforts often have more ambitious goals (Sautter et al. 2007). For example, to maximize the use of the contents, text documents are annotated and indexed according to domain-specific models. The Virtual Itineraries in the

D. Palacio (✉) • C. Sallaberry • M. Gaio
LIUPPA, Université de Pau, avenue de l'Université, Pau 64000, France
e-mail: damien.palacio@univ-pau.fr; christian.sallaberry@univ-pau.fr; mauro.gaio@univ-pau.fr

A. Yeh et al. (eds.), *Advances in Spatial Data Handling and GIS*, Lecture Notes in Geoinformation and Cartography,
DOI 10.1007/978-3-642-25926-5_6,

Pyrenees[1] (PIV) project[2] consists in managing a repository of the electronic versions of books (histories, travelogues) from the nineteenth and twentieth centuries. It appears that the contents present many geographic aspects (Marquesuzaà et al. 2005). This kind of repository is quite stable (few suppressions and modifications, regular inserts of documents) and not too large. Therefore, the cost of a back-office refined semantic aware automated indexing is reasonable (Gaio et al. 2008).

Although well-known search engines still deliver good results for pure keyword searches, it has been observed that precision is decreasing, which in turn means that a user has to spend more time in exploring retrieved documents in order to find those that satisfy his information needs (Kanhabua and Nørvag 2008). One way of improving precision is to include a geographical dimension into the search. We consider the generally accepted hypothesis that Geographical Information (GI) is made up of three components namely spatial, temporal and thematic. A typical textual sample is: "Fortified towns in the south of the Aquitaine basin in the 13th century." To process this textual unit, we claim that each of its three components (spatial, temporal and thematic) should be treated independently, as is put forth by (Clough et al. 2006). This can be done by making several indexes, one per component, as is advised by (Martins et al. 2005). In this way, one can limit the search to one criterion and easily manage the indexes (e.g., to allow adding documents to the corpus). So, our approach consists in processing components independently, in order to better combine them later on. It contributes to the field of Geographic Information Retrieval (GIR) as defined by (Jones and Purves 2006).

The current version of the PIV platform is comprised of three independent process flows: spatial (Gaio et al. 2008), temporal (Le Parc-Lacayrelle et al. 2007) and thematic (Sallaberry et al. 2007). For example, Fig. 6.1 illustrates automatic annotations resulting from such process flows: spatial information is highlighted, temporal information is outlined and the thematic one is underlined. Figure 6.2 illustrates the richness and accuracy of the resulting specific indexes: i.e., the PIV computes geometric representations of spatial information, time intervals corresponding to temporal information and lists of terms corresponding to thematic information. Experiments (Sallaberry et al. 2007) demonstrate the effectiveness of these indexes within specific spatial, temporal or thematic information retrieval scenarios. Two important problems were pointed out during these experiments: 1-results scoring does not integrate spatial features or temporal features frequency within documents: e.g. we are looking for "Biarritz," D1 and D2 will have the same weight even if D1 contains only "Biarritz" spatial feature whereas D2 contains "Biarritz" spatial feature and many other ones; two-merging results within a geographic information retrieval process remains a challenge (Visser 2004): as each index is built with one dedicated approach, as well as each document relevancy calculation formula is based on different methods (which correspond

[1] Mountains of the south west of France.

[2] Part of this project is supported by the Greater Pau City Council and the MIDR media library.

D1: [...] I visited Biarritz during summer 2000. [...]
D2: [...] Wednesday 16th October 2009
[...] The tramway was often out of order during this week in Bordeaux. [...] I plan to leave Bordeaux next week-end and to go to Biarritz. [...] Saturday, a walk near Bayonne, Sunday, a hike at La Rhune peak as well as at Sare.

Fig. 6.1 Example of automatically annotated textual documents

SF_Id	Doc_Id	Text	Geometry
#1	D1	'Biarritz'	(.122...
...			
#5	D2	'near Bayonne'	(.121...
#6	D2	'La Rhune peak'	(.123...

T_Id	Term	Frequency
#1	'visit'	D1,1;
#2	'tramway'	D2,1; D3,1
...		
#7	'walk'	D2,1; D4,2

TF_Id	Doc_Id	Text	Time interval
#1	D1	'during summer 2000'	06/21/2000-09/21/2000
...			

Fig. 6.2 Example of spatial, temporal and thematic indexes

respectively to spatial, temporal or thematic criteria), how to combine spatial, temporal and thematic specific relevancy scores of the retrieved documents?

We propose to normalize each geographic indexing criteria. It consists on rearranging geographic information within a uniform representation form: we represent geographic information within spatial tiles (spatial areas), temporal tiles (calendar intervals) and thematic tiles (concepts) and compute each tile evocation frequency in the documents. Then, we apply statistic formulae generally used for plain-text information retrieval to compute relevancy scores for each resulting document.

This approach proposes (1) frequency parameter integration within the relevance scoring algorithms and (2) geographic data normalization within a new level of spatial, temporal and thematic indexes. Moreover, we propose to produce different granularity level indexes (for example, spatial administrative segmentations: cities, counties, countries) in order to parse the indexes best suited to the grain of each query.

Merging results provided by such hybrid querying criteria would only make sense if such normalized indexes were homogeneous as well as if the relevance calculation formulae were similar. That is why the next section presents a spatial normalization approach, which we will later apply to the temporal and thematic aspects.

The paper is organized as follows. Section 6.2 briefly outlines the textual process flow indexing geographic information within the PIV prototype. Section 6.3 describes related works and our proposals for the creation of new indexes through spatial normalization. Section 6.4 details the proposed model for computing spatial relevance and describes experiments we carried out to evaluate these propositions. Finally, Sects. 6.5 and 6.6 discuss our future perspectives and conclude.

6.2 Textual Process Flow Leading to Spatial Normalization

As explained hereinafter, our data processing sequence is quite different. This spatial information process flow is described in Fig. 6.3. Steps 1–4 are detailed in (Gaio et al. 2008). This approach was developed and experimented within the PIV project:

1. After a classical textual tokenization preprocessing sequence and according to (Baccino and Pynte 1994) we adopt an active reading behavior, that is to say sought-after information is a priori known. A marker of candidate spatial token locates spatial named entities using typographic and lexical rules (involving spatial features initiator lexicons). Then, a morphosyntactic analyzer associates a lemma and a nature with each candidate token (e.g. "Marais", noun).
2. A semantic analyzer marks candidate Absolute Spatial Features (ASF, e.g. "Marais district") first and candidate Relative Spatial Features (RSF, e.g. "Marais district vicinity") next thanks to a Definite Clause Grammar (DCG). For instance, syntagms of composed nouns (i.e. "Marais district," "Emile Zola street,"

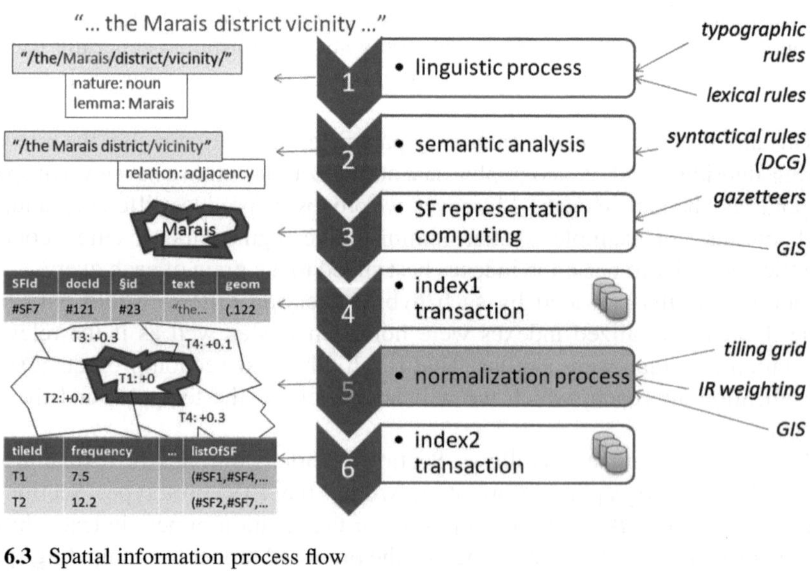

Fig. 6.3 Spatial information process flow

"Wild Chamois peak") are brought together and spatial relationships (adjacency, inclusion, distance, cardinal direction) are tagged (Egenhofer 1991).

3. ASF are validated and geolocalized thanks to external and/or internal gazetteers (IGN French Geographic Institute resources, Geonames resources and contributive hand-craft local resources). Then expressions containing RSF are built from pointed out ASF: embedded spatial relationships (e.g. adjacency: "vicinity") are interpreted and corresponding geometries are computed.
4. Only validated spatial features are retained. Thus we get a spatial index describing each SF with the corresponding geometry, text, paragraph and document. This first level of index supports IR scenarios: query/index overlapping geometries are computed and scored relevant textual paragraphs are returned.
5. SF are grouped, weighted and mapped into a set of segmented grids. We propose different grained tiling grids: regular and administrative grids (district, city, county ...). We use information retrieval TF.IDF formulae (Sparck Jones 1972) to compute spatial tiles' frequencies and weight them.
6. Finally, we get a spatial index describing each tile with the corresponding frequency and SF (geometry, text, paragraph and document). This second level of index supports new IR capabilities: a query is mapped to the more convenient grid and query/index overlapping areas are computed and relevance scoring algorithms integrate each tile frequency. This promotes textual paragraphs centered on the required SF only. Moreover, it allows different querying strategies: for example, thin-scale queries are compared to district grids and large-scale ones are compared to country grids.

A GIS supports spatial operations of all the previous stages. This paper focuses on the spatial information normalization process (stage 5). It describes statistical IR approaches integration in such a process. An experiment compares different index tiling grids and IR statistical formulae to validate our propositions.

6.3 Spatial Information Gathering for Spatial Normalization

6.3.1 Related Works

One of the most popular models developed in textual-based information retrieval research is the vector space model (Salton and McGill 1983). Using a vector space model, the content of each document can be approximately described by a vector of (content-bearing) terms (Cai 2002). An information retrieval system stores a representation of a document collection using a document-by-term matrix (Fig. 6.4), where the element at (i, j) position corresponds to the frequency of occurrence of term i in the jth document (Manning et al. 2008; Cai 2002).

The vector space model can support selecting and ranking of documents by computing a similarity measure between a document and a query or another

Fig. 6.4 Document-by-T matrix within the vector space model

$$
\begin{array}{c} \\ D_1 \\ D_2 \\ \vdots \\ D_n \end{array}
\begin{array}{c}
\begin{array}{cccc} T_1 & T_2 & \dots & T_t \end{array} \\
\begin{pmatrix}
w_{11} & w_{21} & \dots & w_{t1} \\
w_{21} & w_{22} & \dots & w_{t2} \\
\vdots & \vdots & & \vdots \\
w_{n1} & w_{n2} & \dots & w_{tn}
\end{pmatrix}
\end{array}
$$

document (Salton and McGill 1983). There are obvious advantages and disadvantages of using vector space model in retrieving geographical information. Vector space model is well accepted as an effective approach in modeling thematic subspace and it allows spatial information to be handled the same way as thematic information (Cai 2002). Cai (2002) proposed to manage place names within a vector space model. Place names are integrated as independent dimensions in a vector space model, whereas in fact, they are points (or regions) in a two-dimensional geographical space. In order to improve such a keyword-based search method, (Cai 2002) proposed to integrate proper ontologies of places as promoted by (Jones et al. 2001).

Our approach is different as it extends such a term-based matrix to a tile-based matrix. In the vector space model, all the objects (terms, spatial tiles, temporal tiles, thematic tiles (concepts)) can be similarly represented as vectors. This paper proposes to gather SF into spatial tiling grids to compute a similar document-by-tile matrix (Fig. 6.4), where the element at (i, j) position corresponds to the frequency of occurrence of spatial tile i in the jth document.

On the one hand, current spatial oriented research works distinguish:

- Spatial generalization: defined as spatial features selection, displacement and/or simplification processes (Zhang 2005; Zhou and Jones 2004; Zhou et al. 2000; Glander and Dollner 2007);
- Spatial normalization: defined as an image registration process estimating and applying warp-fields (Robbins et al. 2003);
- Spatial summarization: defined as spatial features aggregation/combination into larger features (i.e. cell-based structure) (Rees 2003).

On the other hand, information retrieval oriented research works define normalization as a stemming process of words in order to gather and weight them (Sparck Jones 1972; Li et al. 2002). So, what we call normalizing spatial information, in the following section, means spatial information (representations computed from textual documents) gathering into spatial tiles in order to weight them according to frequency computations.

The originality of the approach described in the following section consists in:

(a) The proposition of different granularity level spatial indexes: administrative and/or regular grids;
(b) The adaptation of effective full-text IR techniques in order to process such indexes.

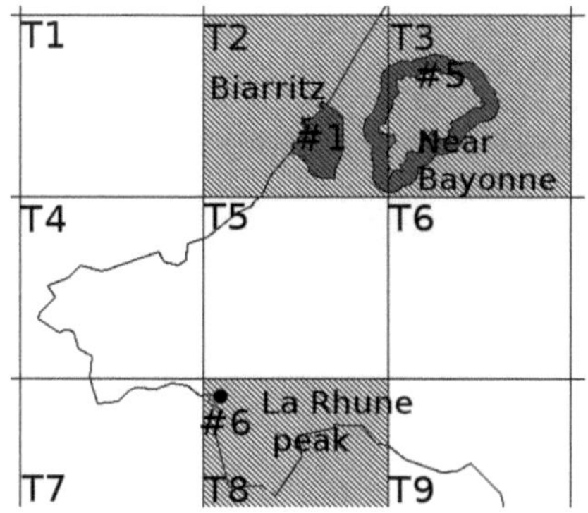

Fig. 6.5 Part of thin-grained SF obtained in index1 projected on a segmentation by regular grid

6.3.2 *Spatial Gathering for Normalization*

First, we detail the spatial normalization process (stage 5 Fig. 6.3) leading to the index2 (stage 6 Fig. 6.3). Then, we briefly explain how we take advantage of this normalized index within an IR process.

6.3.2.1 Information Indexing

Our approach consists in gathering spatial information into a unique type of spatial representation: the tile. So we divide space by attaching each detected SF to tiles. It is similar to the lemmatization process, for which each term is attached to a lemma. Two segmentation types are possible. The first concerns regular tiles (i.e., segmentation into rectangular tiles of the same size—see Fig. 6.5). It is similar to truncation.[3] The second concerns administrative tiles (i.e., segmentation into cities for example—see Fig. 6.6). It is similar to lemmatization.[4] To calculate a tile frequency, one just has to count the number of SF that intersect it, while keeping in mind that a SF can intersect several tiles.

For illustration purposes, let's go back to the example in Figs. 6.1 and 6.2. If we choose to use regular segmentation (Fig. 6.5), we obtain the tiles index shown in Table 6.1. In this table, several scenarios are presented. First, SF #5 intersects two tiles (T2 and T3); so the discrete frequency of both of them is incremented by 1. Moreover tile T2 is intersected by two SF (#1 and #5); consequently it has a weight of 2.

One should note the granularity problem of the spatial information that is being processed, and the proportionality issue between its representation (SF) and a tile as well as the size of their overlapping area. Indeed, one may wonder whether a SF's

[3] E.g. for word "forgotten" the truncation returns "forgot".

[4] E.g. for word "forgotten" the lemmatization returns "forget".

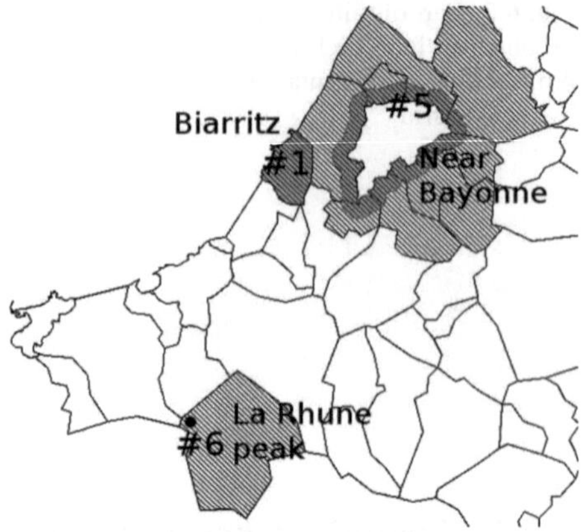

Fig. 6.6 Part of thin-grained SF obtained in index1 projected on an administrative segmentation (cities)

Table 6.1 Spatial index2 with regular tiles (phase 6—Fig. 6.3)

Id_t	Id_{sf} list	Discrete frequency	Continuous frequency
T1	[/]	0	0.00
T2	[#1, #5]	2	0.15
T3	[#5]	1	0.20
...			

Table 6.2 Frequency formulae

Discrete frequency	$freq_t = \sum freq_{sf}$
Continuous frequency	$freq_t = \sum freq_{sf} \times \frac{Ar_{sf,t}}{Ar_t} \times \frac{1}{NbTiles_{sf}}$

$Ar_{sf,t}$ SF area on tile t, Ar_t tile t area, $NbTiles_{sf}$ number of tiles intersected by the SF

area that only covers a small part of a tile should have as great a weight as a SF's area that covers most of the area of the same tile? Thus, we suggest two frequency calculation (see Table 6.2). Indexing may be discrete; so, for a given document unit, a tile frequency is incremented by 1 every time there is an intersection with a SF (Table 6.1 column 3). We have also considered a continuous indexing approach. According to the ratio overlay SF/tile, a tile frequency is incremented by a value between 0 and 1 (Table 6.1 column 4).

6.3.2.2 Information Retrieval

We propose four IR formulae (Table 6.3). TF, TF.IDF and OkapiBM25 are dedicated to discrete weighting. They are widely used in full-text IR (Manning et al. 2008; Savoy 2002). In our context, the TF formula avoids reducing the weight of overly frequent tiles. Nevertheless, classical frequency does not take spatial specificity like granularity into account. That is why we decided to apply TF onto continuous frequency, and we call this approach TFc.

Table 6.3 Weighting formulae, used with index2, for a tile t and a document unit Du

Tile frequency (TF)	$W_{t,Du} = TF_{t,Du} = \frac{freq_{t,Du}}{\sum_{i=1}^{n} freq_i}$
TF.IDF	$W_{t,Du} = TF_{t,Du} \times IDF_t \; and \; IDF_t = log\left(\frac{NDu}{NDu_t}\right)$
OkapiBM25	$W_{t,Du} = \left(\frac{(k_1+1)\times TF_{t,Du}}{(K+TF_{t,Du})}\right) and \; K = k1 \times \left[(1-b) + \frac{b\times n}{advl}\right]$
TF_c	$W_{t,Du} = TFc_{t,Du} = \frac{freqC_{t,Du}}{\sum_{i=1}^{n} freq_i}$

$freq_{t,Du}$ tile frequency in the document unit, n number of tiles in the document unit, NDu_t number of document units with tile t, NDu number of document units, $k_1 = 1.2$, $b = 0.75$, $advl = 900$, $freqC$ continuous frequency

These indexes are intended to support spatial IR. This involves weighting the results. Consequently, we use regular IR formulae and carry out experiments on such indexes of spatial tiles.

6.4 Experiments

Our hypothesis is that the segmentation must be adapted to the SF type of the corpus and of the queries. We propose to use the granularity of the query to choose the best suited index. For complex queries, composed of different grained SF, we propose to define a default well suited index. So here we are looking for what segmentation and weighting formula give the better results for our corpus.

Our approaches are based on thin-grained spatial data. But spatial evaluation campaigns like GeoCLEF (Mandl et al. 2007) do not give accurate resources (like polygons) and do not handle French documents. That's why we realize our experiment on our cultural heritage digital library.

Ten French books were indexed. In order to compare our propositions to manually sorted methods, we chose a sample of 1,019 document units, corresponding to 1,028 SF (902 ASF and 126 RSF). Each document unit may contain from 0 to many SF. We submitted 40 queries (the baseline is index1). Fifteen queries involve an ASF: five of each type (small grained like peaks, intermediate grained like cities and large grained like regions). Twenty five other contain a RSF: five of each type (orientation, proximity, union, inclusion, distance). Table 6.4 shows some examples of relations. We observed that 30% of our ASF are well identified cities, 12% are larger well identified ASF (department, regions), 38% are smaller ASF (peaks, cabins ...) and the others (approximately 20%) have a variable average size.

We tried six different indexing segmentations: three administrative segmentations (city, department and region) and three regular segmentations (grid of 100 $\times$ 100, grid of 200 $\times$ 200 and grid of 400 $\times$ 400). The grid of 200 $\times$ 200 corresponds to the average city size. Finally, we tested all theses segmentations with the 4 weighting formulae presented in last section (TF, TF-IDF, OkapiBM25, TFc).

Table 6.4 Examples of RSF

In downtown Paris (inclusion)	Near Gavarnie (proximity)
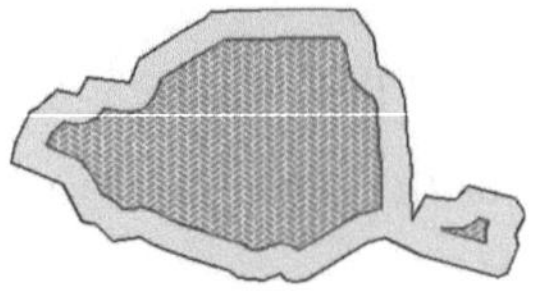	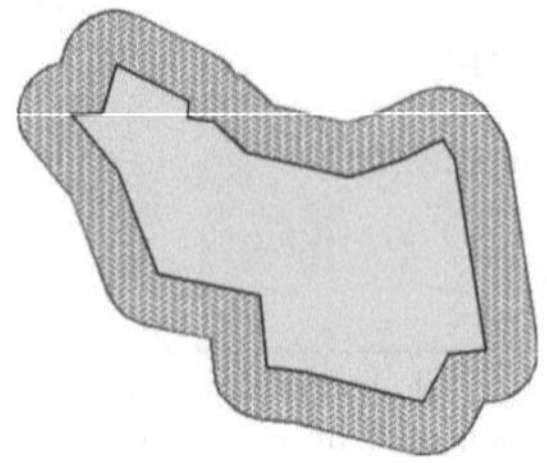
On Tarbes-Lourdes axis (union)	In south of Ile-de-France (orientation)
	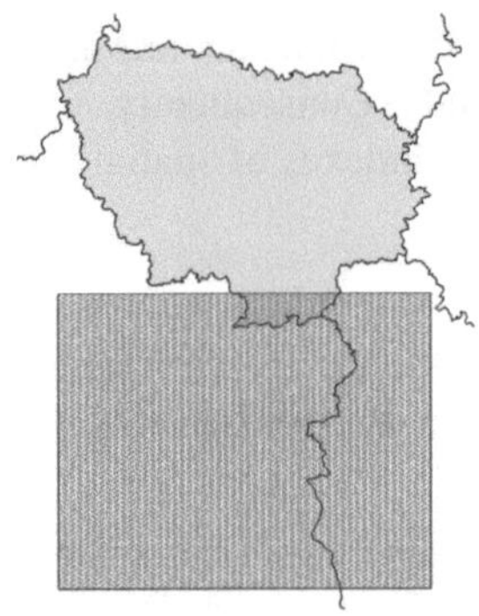

As we can see in Table 6.5, for all segmentations, the TFc gives the best results. As we explained in Sect. 6.3, every classical statistical weighting formulae (TF, TF-IDF, OkapiBM25) use discrete frequency. They give the same weight for a geometry which fills the major part of one tile, and for a geometry which represents a little part of the same tile. On the contrary, the TFc uses continuous frequency and gives a weight depending on the area of overlapping between the tile and the SF's geometry.

Concerning the segmentation, the Table 6.5 shows that all segmentations give good results excepts department and regions (they are too large so they gather SF which are too far away from each other). For segmentation by regular grid, the one of 200 $\times$ 200 gives the best results. Concerning the administrative segmentation, city segmentation gives the best results. The main explanation is that an important part of the indexed ASF concerns well identified cities. So it confirms our hypothesis that the segmentation must be adapted to the type of the SF contained in the corpus.

Tables 6.5 and 6.6 also show that the city segmentation associated to the TFc gives better results than our baseline. Let's take example of Fig. 6.1 to illustrate why we obtain such results. If we consider query "in Biarritz," the relevancy score for D2 on index1 is 1.0 because the text contains the SF "in Biarritz." It does not

Table 6.5 Results of experiment on SF with different segmentations and weighting formulae

MAP	TF	TF-IDF	Okapi	TFc
City segmentation	0.61	0.61	0.63	**0.70**
Department segmentation	0.40	0.39	0.40	**0.53**
Region segmentation	0.40	0.39	0.39	**0.56**
Grid of 100 × 100	0.59	0.59	0.62	**0.68**
Grid of 200 × 200	0.61	0.60	0.63	**0.69**
Grid of 400 × 400	0.63	0.62	0.65	**0.66**

Table 6.6 Results of experiment on SF with baseline (index1)

MAP	Spatial overlapping
Index1 (baseline)	0.62

take into account the other SF. On the other hand, the city segmentation associated to the TFc gives a relevancy score of about 0.17. It computes a lower score to the document unit because it contains other less relevant SF.

In conclusion, we advise segmentation into cities and the TFc formula (cf Table 6.5) for cultural heritage digital libraries. This normalization allows one to introduce an initial approximation of the spatial context (weighting a document unit takes into account all the SF it contains).

6.5 Ongoing and Future Work

The PIV platform supports a similar processing sequence producing temporal indexes (Fig. 6.3). It deals with calendar temporal features (CTF) that may be absolute (ACTF) or relative (RCTF) like spatial ones.

Let one consider that the previous text sample involves the following temporal features: CTF1—"the 26th of December", CTF2—"Saturday 29th of December at 2 pm", CTF3—"at the beginning of the winter", CTF4—"the last days of December 1933". The PIV produces such an index (Table 6.7, Fig. 6.7).

The PIV temporal information normalization process (ongoing development similar to the spatial normalization process) would return weighted temporal intervals presented in Fig. 6.8. This example illustrates calendar segmentation where each interval represents a week: CTF4 intersects weeks W51 and W52. For example the week W52 has a weight of 4 according to the discrete indexing approach.

Currently, we are working on temporal normalization experiment. We aim to propose spatial and temporal criteria combination strategies with geographic IR scenario.

Table 6.7 Index of temporal features in PIV

Id_{cft}	Text	Type	Timestamp
ctf1	The 26th of December	actf	(1933-12-26, 1933-12-26)
ctf2	Saturday 29th of December	actf	(1933-12-29, 1933-12-29)
ctf3	At the beginning of the winter	rctf	(1933-12-22, 1934-03-19)
ctf4	The last days of December 1933	rctf	(1933-12-21, 1933-12-31)

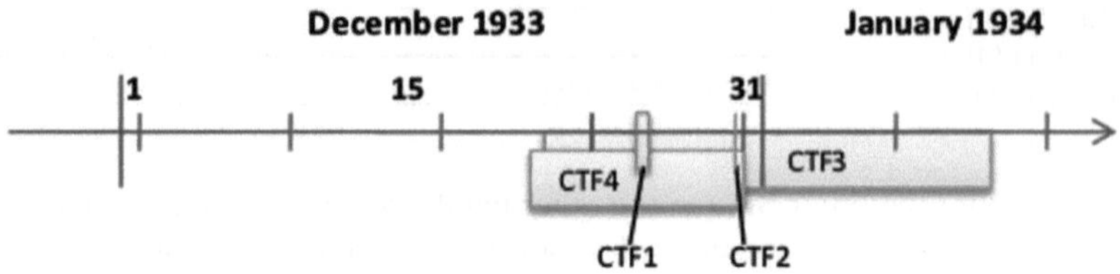

Fig. 6.7 Temporal index

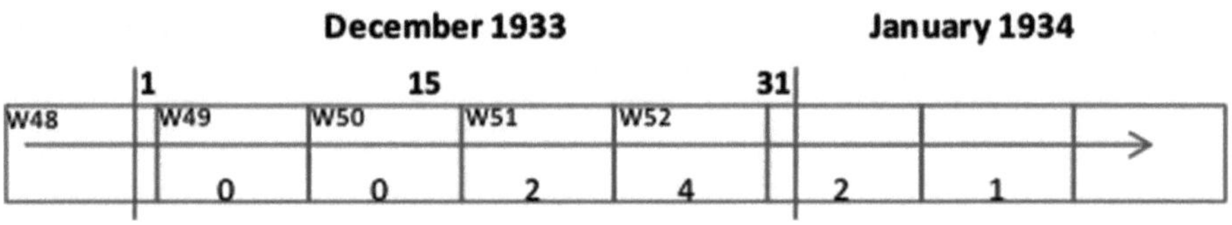

Fig. 6.8 Calendar segmentation

6.6 Conclusion

The Virtual Itineraries in the Pyrenees (PIV) project consists in managing a repository of the electronic versions of books (histories, travelogues) from the nineteenth and twentieth centuries. The PIV engine automatically annotates, interprets and indexes spatial, temporal and thematic information contained in those documents. Three independent process flows support spatial, temporal and thematic indexing and IR operations.

Two important problems were pointed out during a first campaign of experiments (Sallaberry et al. 2007): one-results scoring does not integrate spatial or temporal features frequency within documents; two-merging results within a geographic information retrieval process remains a challenge. The main problem of current geographic IR systems comes from the fact that the index structure and relevancy computation approaches used for space, time and theme are intrinsically different (Visser 2004).

Our hypothesis is based on a spatial, temporal and thematic tiling of information in order to build higher level indexes and to adapt effective full-text IR techniques to process such indexes.

In this paper we propose an approach for normalizing spatial indexes automatically. Such a gathering of spatial features into spatial tiles implies some loss of accuracy. However, as we have different grained indexes, we may select the best suited one during a querying stage. Moreover, experiments point out the effectiveness of our solution: a continuous spatial tiles frequency computation associated to a continuous document units relevancy computation formula gives better results than our baseline dedicated to the weighting of the most relevant SF of a document unit.

As explained before, the aim of this normalization method is to develop a general indexing strategy that is suited for spatial, temporal and thematic information in order to combine such geographic IR results. We are currently working on the evaluation of the effectiveness of this indexation on a larger sample of texts and queries. We are also working to apply normalization methods for the building of normalized temporal and thematic indexes from textual input. Future improvement of the presented approach would be to explore how to combine normalized spatial, temporal and thematic indexes and compute a unique relevancy scoring. Merging results for a geographic IR approach combining such different criteria is a recurring research question nowadays (Martins et al. 2008; Vaid et al. 2005).

References

Baccino T, Pynte J (1994) Spatial coding and discourse models during text reading. Lang Cogn Process 9:143–155

Cai G (2002) GeoVSM: an integrated retrieval model for geographic information. In: Egenhofer MJ, Mark DM (eds) GIScience. Lecture notes in computer science, vol 2478. Springer, Boulder, CO, USA, pp 65–79

Clough P, Joho H, Purves R (2006) Judging the spatial relevance of documents for GIR. In: ECIR'06: Proceedings of the 28th European conference on IR research, April 2006, Lecture notes in computer science, vol 3936. Springer, London, UK, pp 548–552

Egenhofer MJ (1991) Reasoning about Binary Topological Relations. In: Gunther O, Schek H-J (eds) SSD. Lecture notes in computer science, vol 525. Springer, Zürich, Switzerland, pp 143–160

Gaio M, Sallaberry C, Etcheverry P, Marquesuzaa C, Lesbegueries J (2008) A global process to access documents' contents from a geographical point of view. J Vis Lang Comput 19(1):3–23

Glander T, Dollner J (2007) Cell-based generalization of 3D building groups with outlier management. In: Samet H, Shahabi C, Schneider M (eds) GIS. ACM, Seattle, WA, USA, p 54

Jones CB, Purves R (2006) GIR'05 2005 ACM workshop on geographical information retrieval. SIGIR Forum 40(1):34–37

Jones CB, Alani H, Tudhope D (2001) Geographical information retrieval with ontologies of place. In: Montello DR (ed) Proceedings of the conference on spatial information theory (COSIT 2001). Lecture notes in computer science, vol 2205. Springer, Heidelberg/Morro Bayand, pp 322–335

Kanhabua N, Nørvag K (2008) Improving temporal language models for determining time of non-timestamped documents. In: ECDL'08: Proceedings of the 12th European conference on research and advanced technology for digital libraries, Springer, Berlin/Heidelberg, pp 358–370

Le Parc-Lacayrelle A, Gaio M, Sallaberry C (2007) La composante temps dans l'information géographique textuelle. Revue Document Numérique 10(2):129–148

Li H, Srihari KR, Niu C, Li W (2002) Location normalization for information extraction. In: 19th international conference on computational linguistics (COLING 2002). Howard International House and Academia Sinica, Taipei, Association for Computational Linguistics

Mandl T, Gey FC, Nunzio GMD, Ferro N, Larson R, Sanderson M, Santos D, Womser-Hacker C, Xie X (2007) GeoCLEF 2007: the CLEF 2007 cross-language geographic information retrieval track overview. In: Peters C, Jijkoun V, Mandl T, Muller H, Oard DW, Penas A, Petras V, Santos D (eds) CLEF. Lecture notes in computer science, vol 5152. Springer, Budapest, Hungary, pp 745–772

Manning CD, Raghavan P, Schutze H (2008) Introduction to information retrieval. Cambridge University Press, New York

Marquesuzaà C, Etcheverry P, Lesbegueries J (2005) Exploiting geospatial markers to explore and resocialize localized documents. In: Rodriguez MA, Cruz IF, Egenhofer MJ, Levashkin S (eds) GeoS. Lecture notes in computer science, vol 3799. Springer, Mexico City, Mexico, pp 153–165

Martins B, Silva MJ, Andrade L (2005) Indexing and ranking in Geo-IR systems. In: GIR'05: Proceedings of the 2005 workshop on geographic information retrieval, ACM, New York, pp 31–34

Martins B, Manguinhas H, Borbinha JL (2008) Extracting and exploring the geo-temporal semantics of textual resources. In: Proceedings of the IEEE international conference on semantic computing. (ICSC'08), IEEE Computer Society, Washington, DC, USA, pp 1–9

Rees T (2003) "C-squares", a new spatial indexing system and its applicability to the description of oceanographic datasets. Oceanography 16(1):11–19

Robbins S, Evans AC, Collins DL, Whitesides S (2003) Tuning and comparing spatial normalization methods. In: Ellis RE, Peters TM (eds) MICCAI (2). Lecture notes in computer science, vol 2879. Springer, Montréal, Canada, pp 910–917

Sallaberry C, Baziz M, Lesbegueries J, Gaio M (2007) Towards an IE and IR system dealing with spatial information in digital libraries – evaluation case study. In: ICEIS'07: Proceedings of the 9th international sonference on enterprise information systems, Funchal, Madeira, Portugal, pp 190–197

Salton G, McGill MJ (1983) Introduction to modern information retrieval. McGraw-Hill, New York, NY, USA

Sautter G, Bohm K, Padberg F, Tichy WF (2007) Empirical evaluation of semi-automated XML annotation of text documents with the GoldenGATE Editor. In: ECDL'07: Proceedings of the 11th European conference on digital libraries. Lecture notes in computer science, vol 4675. Springer, Budapest, Hungary, pp 357–367

Savoy J (2002) Morphologie et recherche d'information. Technical report, Institut interfacultaire d'informatique, Université de Neuchatel, Neuchatel

Sparck Jones K (1972) A statistical interpretation of term specificity and its application in retrieval. J Docum 28(1):11–21

Vaid S, Jones CB, Joho H, Sanderson M (2005) Spatio-textual indexing for geographical search on the web. In: Medeiros CB, Egenhofer MJ, Bertino E (eds) SSTD. Lecture notes in computer science, vol 3633. Springer, Angra dos Reis, Brazil, pp 218–235

Visser U (2004) Intelligent information integration for the semantic web. Springer, Heidelberg

Zhang Q (2005) Road network generalization based on connection analysis. In: Developments in spatial data handling. Springer, Berlin/Heidelberg, pp 343–353

Zhou S, Jones CB (2004) Shape-aware line generalisation with weighted effective area. In: Fisher PF (ed) Developments in spatial data handling 11th international symposium on spatial data handling. Springer, Kyoto, Japan, pp 369–380

Zhou X, Zhang Y, Lu S, Chen G (2000) On spatial information retrieval and database generalization. In: Proceedings of the Kyoto international conference on digital libraries. Kyoto, pp 380–386

Chapter 7
Explore Multivariable Spatio-Temporal Data with the Time Wave: Case Study on Meteorological Data

Xia Li and Menno-Jan Kraak

Abstract Traditionally the GIScience community is well able to deal with the locational and attribute component of spatio-temporal data. However, the methods and techniques to deal with the data's temporal component are less developed. This paper introduces a conceptual framework that combines user tasks, available temporal data and visualization theories to discuss temporal visualization. Two limitations of existing method are improved by introducing the time wave environment which is a close combination of temporal graphic representation and temporal interactive tools, and operates in so-called time space. Time space is a concept which represents time and answers the temporal questions. This term is based on the data components and user task framework: location (where), attribute (what) and time (when). From a visualization perspective this translates in: location space (maps), attribute space (diagrams) and time space. In this research, the time wave which is a main element in the time space is a good example of how an alternative view on the data might reveal patterns not always obvious from traditional graphic representations. A case study based on meteorological data illustrates this approach.

X. Li (✉)
College of Earth Science and Resources, Chang'an University, 126# YanTa Road, Xi'an City 710054, China
Institute of Hydrogeology and Environment Geology of the Chinese Academy of Geological Sciences
e-mail: lixia666@chd.edu.cn

M.-J. Kraak
Faculty of Geoinformation Science and Earth Observation, University of Twente, Enschede, The Netherlands

A. Yeh et al. (eds.), *Advances in Spatial Data Handling and GIS*,
Lecture Notes in Geoinformation and Cartography,
DOI 10.1007/978-3-642-25926-5_7,

7.1 Introduction

Many of the most important challenges our society is facing today, such as global climate change, economic development and infectious diseases depend on spatio-temporal data to detect and analyze changes as well as trends to support problem solving. Especially the temporal component of the data should be studied carefully to understand the changes and their impacts. Current data collection techniques offer a wide variety of thematic data in many different spatial and temporal resolutions. From a temporal perspective, earth observation techniques provide data with temporal resolutions varying from weeks, days, hours, to even minutes. The challenge faced is how to process, manage, and use these continuous streams of data to support problem solving and decision making. The application of graphic representations in a dynamic and interactive geovisualization environment is part of the solution.

Geovisualization integrates approaches from disciplines including cartography with those from scientific visualization, image analysis, information visualization, exploratory data analysis and GIScience (Dykes et al. 2005). The graphic representations, mostly maps, are used to stimulate (visual) thinking about geospatial patterns, relationships and trends. This is strengthened by looking at the data in a number of alternative ways. Playing with the data using these multiple representations without constraints (traditions) will trigger the mind of the users, and lead to an increase in their knowledge. This does improve our understanding of how to use visualization to get a better insight into spatial data, but not necessarily in temporal data. This is partly due to the fact that most methods and techniques used to solve the geo-problems are from either location or attribute perspective. This paper will look at the geo-problems from a temporal angle, from what is called time space (Li and Kraak 2008). Time space is a visualization space to represent time and answer the temporal questions; for example, what is the temporal distribution of precipitation per month, season or year? How is it linked to location space (e.g. maps) and attributes space (e.g. diagrams).

In the wider context of GIScience research on temporal data analysis, modelling (Goralwalla et al. 1998) and temporal visualization are receiving increased attention (Mackinlay et al. 1991; Harrison et al. 1994; Allen 1995; Brown 1998; Harris et al. 2000). This includes papers that offer an overview over existing visualization methods (Andrienko et al. 2003a, b; Aigner et al. 2007). Most of these focus on applications in the information visualization field or on discussions of specific aspects of temporal visualization such as static and dynamic representations (Muller and Schumann 2003), linear time (Silva and Catarci 2000) and the visualization process (Chi et al. 1998). Others take a more overall approach. Aigner (Aigner, Miksch et al.) discussed temporal visualization based on a time-oriented framework. Andrienko (Andrienko, Andrienko et al.) focused on spatio-temporal exploration and considers the time visualization from both data and user tasks. However, compared with location space, and attribute space, time space hasn't been studied very well.

In this paper, a proposed solution is to take a closer look at the integration of several visualization theories accepted in GIScience and information visualization, and see how to extend these with a specific temporal component. This is done by analyzing user tasks to structure the temporal data. Hereby the limitations of existing temporal visualization method will be discussed and a specific graphic representation will be suggested. To structure time space, the suggested graphic representation should allow the user to represent the temporal data, to display different views on time (e.g. linear or cyclic), and last but not least should allow for interaction. For this the time wave was introduced (Li and Kraak 2008). In the next section, a commonly accepted visual problem solving approach is discussed with specific attention for the nature of the data, the user task and the visualization environment. After this background information a temporal conceptual framework for visualization is presented, followed by a case study that demonstrates the alternative approach.

7.2 Visualization Theory and Time Space

7.2.1 Visualization Theory

A common approach to support problem solving with visualization is shown in the scheme in Fig. 7.1. A set of tasks, translated into questions is executed in an appropriate visualization environment with suitable data to solve the problem at

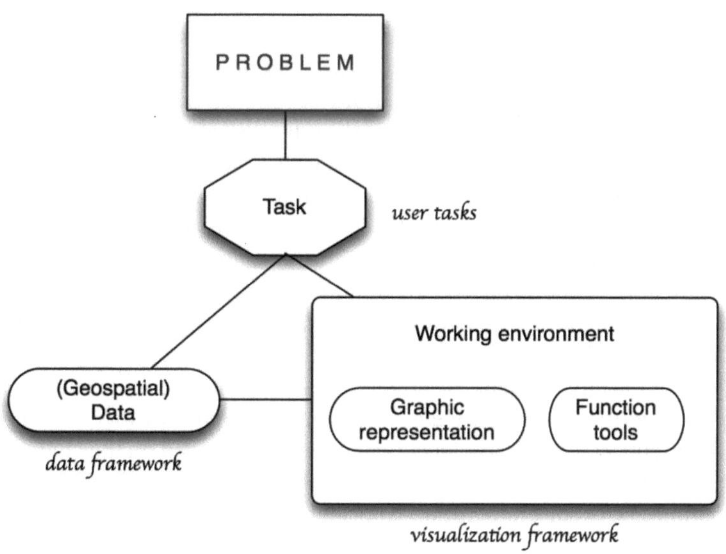

Fig. 7.1 Visual problem solving: the relation between user tasks, a data framework and a visualization framework

hand. Three keywords are of importance: user tasks, data framework and visualization framework. The visualization framework includes the graphic representation and the functional tools to 'play' with the graphics.

Relevant questions are: How to abstract questions based on the data? How to address the data in a visualization environment? How to decide on the right graphic representation(s) and the required functions to answer the questions? Most cartographic and geovisualization theories are studied by looking at the interrelationship of these tightly coupled aspects, such as Bertin (1983), Macrachren (1995), Peuquet (2002) and Andrienko and Andrienko (2006).

7.2.2 *Time Space*

Time space includes temporal graphic representations and temporal interactive tools. Here the requirements of time space will be discussed from temporal data and temporal user tasks perspectives.

7.2.2.1 Temporal Data and Temporal Visualization

Compared with "attribute" and "location", "time" is considered special. Attribute and location can change over time, but the reverse is not possible. The notion of time is difficult to grasp. This has led to many different views on time. Examples relevant in GISciense are found in Frank(1994), Goralwalla et al. (1998), Peuquet (2002) and Aigner et al. (2007). It can be said time passes continuously, but it is observed and measured in discrete points or intervals. This results in a view of continuous time versus discrete time. Time can also be considered as relative (e.g. last week) or absolute (e.g. May 27th). The continuous nature of time makes it linear (before, after) but it obviously holds cyclic characteristics as well (day, season). The granularity of time is defined by the different units in which time is expressed, and can be used to define the scale of time. These granularities inherit from both linear and cyclic characteristics (hours, seasons).

From the temporal data aspect, the views on time such as linear or cyclic, continuous or discrete, single scale or multiple scale, should be considered and addressed in time space properly. In relation to the task each view might have different requirements for temporal visualization. For example, the question 'how did the city expand?' could be answered with discrete multiple views showing a set of snapshots of the city's extent. Alternatively an animation could be used to answer the same question in continues view. The dynamic nature of the animation could also attract attention to other aspects of the expansion. Based on the character of the data, a linear or cyclic representation might by appropriate. Both could express either continuous or discrete time. Although there might be a straightforward solution in the selection process of the visual representation, it could be useful to try other visualizations as well, because alternative view might reveal patterns or aspects of the data that could remain hidden in the straightforward solution.

7.2.2.2 Temporal User Tasks and Temporal Visualization

Even though the selection of a graphic representation is based upon the nature of the data, the user task plays an important role as well, since at the end, the graphic representation has to answer questions. For instance, the question 'When did countries get their independence?' might require different representations depending on the focus of the question. A timeline with names and years along will do if the temporal distribution is the focus. If one is also interested in the spatial distribution then a world map with labels indicating the year of independence could the most suitable representation. If both are of interest, a space-time-cube (Hägerstrand 1967) could be a possible solution. When one would also be interested in the season of independence the graphic representation should be able to handle both linear and cyclic time and the time wave might be a solution.

Shneiderman's visual information seeking mantra is widely accepted in the visualization field and can be applied in time space. First, the temporal graphic representation is used to locate the time, show the temporal overview, and then temporal interactive tools are required to carry out temporal zooming and temporal filtering options to get the details-on-demand. This process is often an interactive and iterative process, which is supported by a close combination of representation and interactive tools in time space. However, this close combination has not been realized in temporal visualization. There are many temporal graphic representations which could display a temporal pattern of data, such as ThemeRiver (Havre et al. 2000), Stacked bar chat (Harris 1999), People garden (Xiong and Donath 1999), MultiCombs (Tominski et al. 2003) and etc. To assist answering the questions, temporal interactive tools should allow one to identify single or multiple points in time, identify points at periodic intervals, or define intervals of certain length in both the linear and the cyclic format. Some have experience with this approach, like Koike et al. (1997) with TimeSlider and Edsall et al. (1997) experimented with a time wheel query device in the TEMPEST system. Above temporal visualization methods are either a temporal representation or a temporal interactive tool only.

7.3 Temporal Visualization and Time Wave: As a New Approach in Time Space

Most temporal representation are either controlled by a timeline (linear time) (Mackinlay et al. 1991; Plaisant et al. 1996; Silva and Catarci 2000; Wijk 2002) or a time wheel (cyclic time) (Mackinlay et al. 1994; Carlis and Konstan 1998; Harris 1999; Daassi et al. 2002). However, this does not always result in satisfying solutions because many phenomena have both linear and cyclic characteristics. The time wave (Fig. 7.2; Li and Kraak 2008) is one potential solution for representing both the linear and cyclic nature of data. It is a combination of the timeline and the time wheel and offers an alternative view by its own. With the timeline and the time wheel it is difficult to show multiple time scales. However, the time wave can show

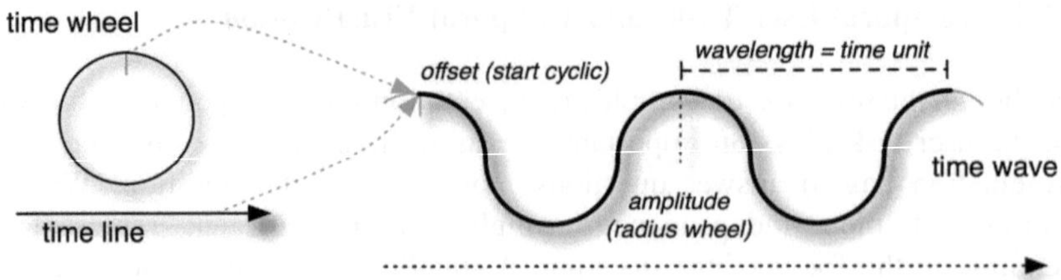

Fig. 7.2 The time wave as a construct of the time line and the time wheel combining the linear and cyclic nature of time (Li and Kraak 2008)

multiple time scales by nesting different waves with different wavelengths and amplitudes based on their temporal scale.

Time wave not only could show the linear and cyclic temporal pattern of data as a temporal representation, but also carry out zooming and filter as an interactive tools. For temporal interactive tools, the time wave can do both linear and cyclic operation as well. In addition, the interactive time wave also support one to represent some locational and attribute characteristics in relation with the phenomenon's temporal distribution which can be explored.

Furthermore, the time wave allows users to interact with the temporal reference or temporal representation. Supported by Coordinated Multiple View (CMV) (Roberts 2005, 2008) principle, one could identify or select any time instant or interval at various scales, or select a temporal pattern, to show the corresponding information in the location space or attribute space. Figure 7.3 shows an interpretation of time space with the time wave combining the representation and manipulation functions to carry out the interactive and iterative exploration process. The figure also demonstrates the link between time space, location space and attribute space. With a temporal question, one starts in time space and depending on the nature of the question, the answer can be given in time space itself or one 'jumps' out of time space into location or attribute space. For instance, the question related to year of independence of the countries is an example where one moves to location space when one needs to see spatial patterns next to temporal patterns. In answering complex question one might have to jump from space to space in an extensive iterative process to identify and compare temporal, spatial and attributer patterns.

7.4 Case Study: Time Space and Meteorological Data

To see how to create a proper time graph based on the analysis of temporal data, user tasks, a case study is discussed in this section. The data set contains observations of 19 meteorological stations in Beijing, China. For each station, known at location (x, y, z), the temperature and the dominant land use information

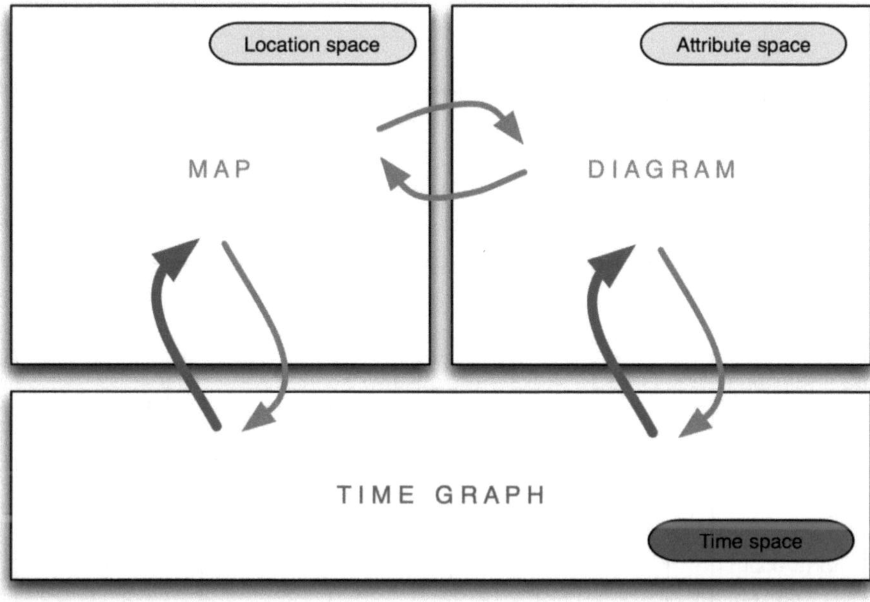

Fig. 7.3 The interactive options to move from time space into location and attribute space supported by CMV principle

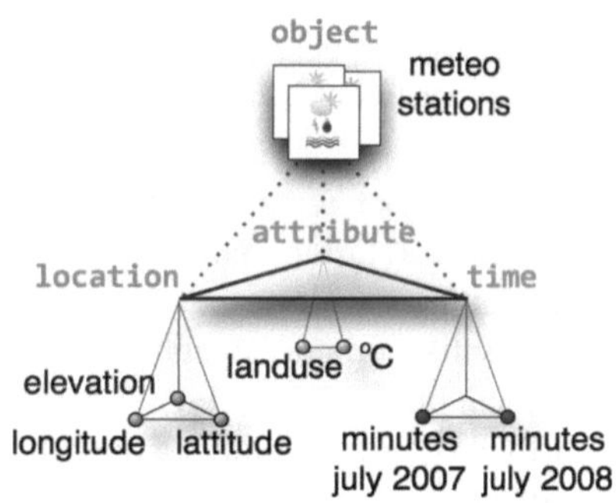

Fig. 7.4 Case study data: the characteristics of the meteorological stations in the Beijing area

are given. The temperature was measured at minute interval over the months in July 2007 and July 2008. A data overview is shown in Fig. 7.4. The location data is point data in context of geographical units (districts in Beijing city); the attribute data is quantitative (temperature) and qualitative (land use); the temporal data has linear characteristics (measurements over a month) and multi-scale cyclic characteristics (July for 2 year and days and nights for 24 h periods).

The user task is to find out when the temperature reaches its daily maximum value at the different meteorological stations. Is the pattern related to the longitude, latitude and elevation of each meteorological station? What are the other impact factors, such as land use? Is there any difference between those 2 years? The temporal questions, such as when, how often, in what order, synchronization and

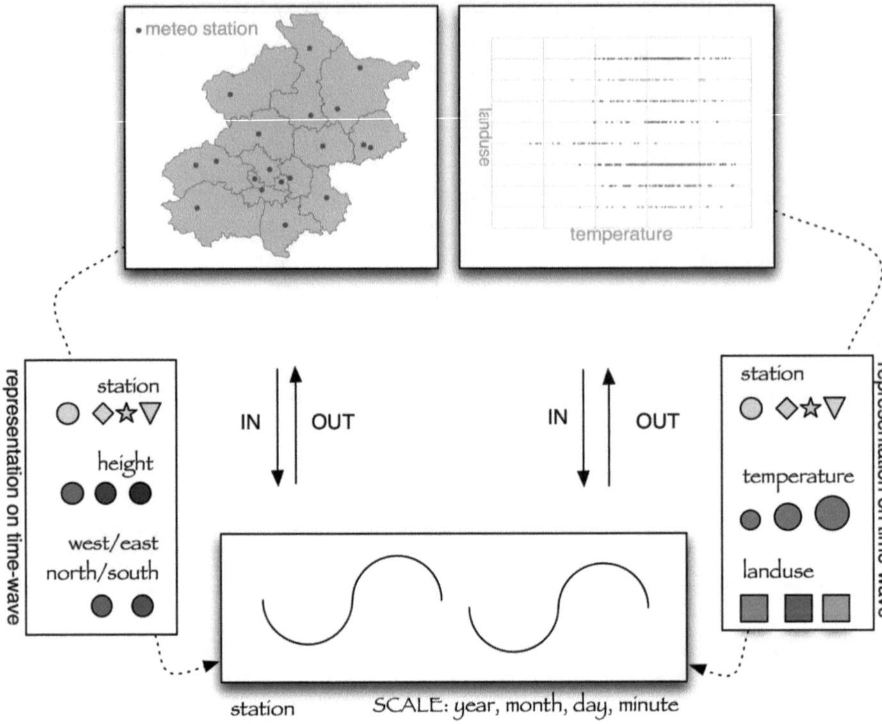

Fig. 7.5 The working environment of the time wave, with options to visually represent characteristics from location and attribute space on the wave

trends, always have a relation to other data components. For instance, from object perspective (station name, station ID), location perspective (height; east/west; north/south), attribute perspective (low, high, or average temperature; land use category like urban, vegetation, water, or other).

Based on Fig. 7.4, Fig. 7.5 gives a schematic overview of the coordinated multiple view environment in which time space plays a key role. The time wave is selected as graphic representation since it can handle the temporal data requirements of the case study: linear and cyclic multi scale time. The wave length could be months, weeks, days or any other time unit one would need. It also allows interaction and is directly linked to location and attribute space which is needed if one has to switch spaces while executing the user tasks. In Fig. 7.5 this is represented by the in and out arrows. Location space show the map with the position of the meteorological stations and attribute space shows a scatterplot of the station's temperature versus land use. Figure 7.6 gives a snapshot of the actual software. The time wave is a plug-in for the Udig open source GIS. The time wave can also include a visual representation of the data's attribute and location characteristics. In Fig. 7.5 symbols related to the meteorological stations, the

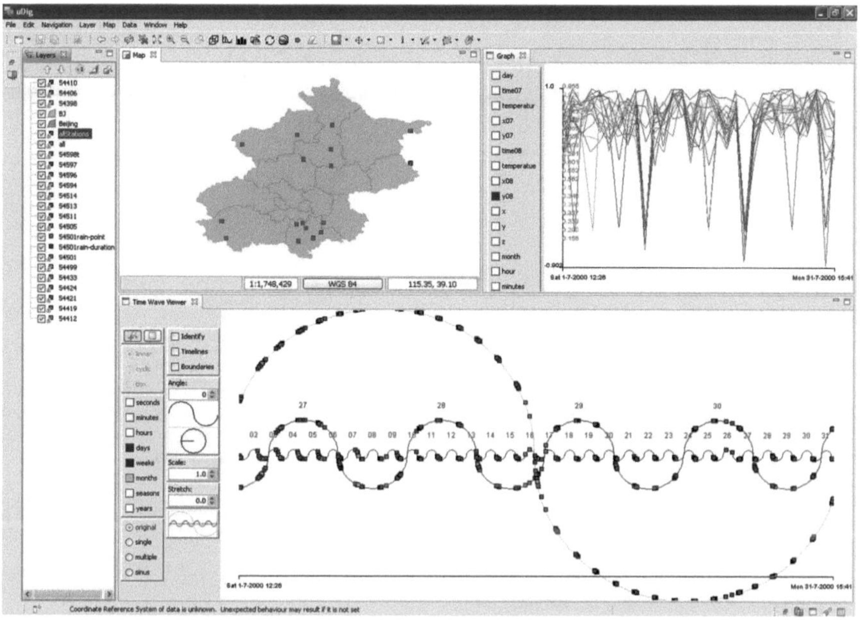

Fig. 7.6 Snapshot of the time space/time wave environment as implemented in the open source GIS uDig

temperatures observed, as well as land use could be plotted on the wave. Figures 7.7 and 7.8 give examples related to the case study.

Figure 7.7 gives two examples of how time space works in practice. In the upper section of the Figure an overview of the month July 2007 is given. The wave length represents a day. For each day and each meteorological station, a symbol is located on the wave at the time when the highest temperature was observed. It is obvious that a wave like this will only offer and overview. A first glance at the data shows that the highest temperature is reached a few hours after 12 noon (the top of the wave). However, both at the beginning and the end of the month an anomaly in the temporal pattern can be observed. These two anomalies have been further explored in the Figure. In both cases one can first zoom in the wave to see more details. At the beginning of the month (the left of the figure) two stations are rather late in reaching their maximum temperature and at the end of the month (the right of the figure) there is a single day on which all stations are late in reaching their maximum temperature. In the first situation it makes sense to get more details on those two particular stations, and switch to location space to see if their location is special. The map doesn't reveal anything special, since stations with similar geographic condition can be found. For the second situation it was decided to switch to attribute space and have a closer look at the temperature ranges of all stations at that particular day and the days before and after. It can be observed that the maximum temperature for that particular day was also considerably lower than the days before

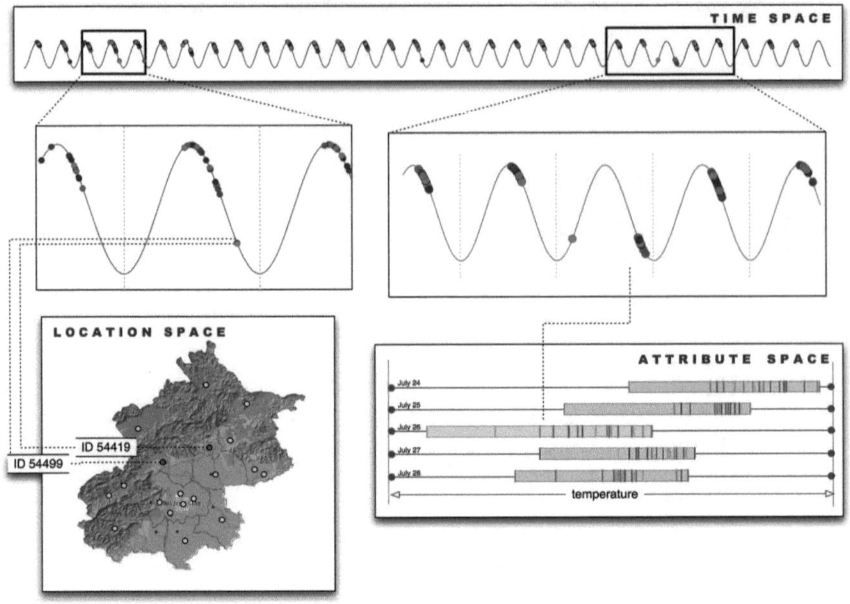

Fig. 7.7 Working with the time wave: identify and explain temporal patterns. At the *top* the overview of July 2007 showing all stations at the moment these observe the highest temperature. In the *middle* zooming in on two clear anomalies in the monthly temporal pattern. At the *lower left* a map in location space and the *lower right* a temperature diagram in attribute space

and after. The available data do not give a good explanation. Other meteorological parameters, for example cloud cover, may be introduced to give an explanation. The experts will have to continue exploring the data and perhaps obtain additional information in order to be able to explain the anomalies.

Figure 7.8 shows another useful function of the time wave. The top wave is the normal time wave, here showing again the moment that the daily temperature reaches its maximum at each meteorological station for 3 days. It is possible to create a set of parallel waves (Fig. 7.8b). For each meteorological station a separate wave is created to obtain a better view on the patterns. This approach has been derived from the parallel coordinate plot. The wave in Fig. 7.8c shows the data sorted based on the values of July 1st. This result in having the station with the earlier highest temperature in the lower wave. Keeping this sort order the following days show a different pattern. It means the order of each stations reach its maximum temperature varies over different day. It is possible to sort the data based on any other available variables. For instance, one could sort based on the height of the stations (Fig. 7.8d), or sort from east to west or north to south. After sorting, one might observe patterns that will stimulate one to jump to location or attribute space, to look for more detail, to switch back to overview mode or even to retrieve more data. Working with the time wave to explore temporal patterns is clearly an iterative

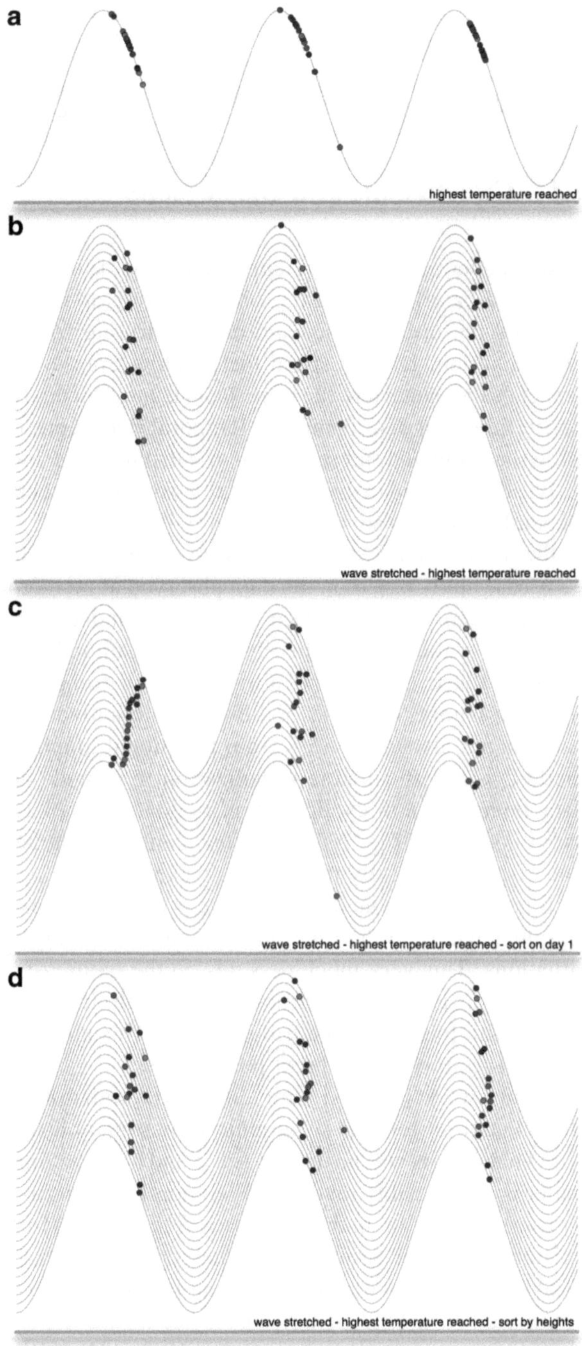

Fig. 7.8 Working with the time wave: comparing temporal patterns. (**a**) an overview of the first 3 days of July 2008 showing when the maximum temperature was observed for each station; (**b**) temporal stretch by creating a set of parallel waves to un-clutter the pattern as seen in the upper wave; (**c**) sorted wave based on the temperature values of July 1, 2008; (**d**) sorted by station's elevation (highest elevation in lowest parallel wave)

process. In the next section the specific advantages and possible limits of the approach will be discussed.

7.5 Conclusions and Discussions

In this paper an alternative approach to work with spatio-temporal data has been presented based on a visual problem solving approach. The nature of the temporal data and the user task at hand are the driving forces to select a suitable graphic representation. In time space the graphic representation focuses on the temporal aspect of the data. Here the time wave is introduced. It not only combines linear and cyclic time, but also combines temporal data representations and interaction, and allows a limited representation of attribute and location data. The exploratory activities are guided by the visual information seeking mantra of Shneiderman. Depending on the user task, one can switch from one space to the other, and into any of the information seeking modes (overview, zoom/filter, and details on demand).

A case study based on the data observed at meteorological station in Beijing has demonstrated the time space framework and the capabilities of the time wave. Temporal patterns and distribution are studied from time space switching to location space (maps) and attribute space (diagrams) when required. The time wave is a good example of how an alternative view on the data might reveal patterns not always obvious from traditional graphic representations. In overview mode it is a very suitable visualization to provide an impression of the nature in data, being linear or cyclic. In this mode the wave easily reveals anomalies as well, as for instance in Fig. 7.7. This effect can be strengthened by moving the horizontal line (x-axis) vertically through the wave.

It is not argued that the time wave is the only visual representation in time space. Graphics based on the time line or time wheel only, would be preferred in certain situations. This very much depends on the nature of the data and the task at hand. However, it is claimed that the optimal solution space is time space interactively linked with location and attribute space. This allows the user to flexibly tackle the problem from many different perspectives.

Future work will deal with a detailed validation of the time wave in other case studies and its relation to graphic representation in location and attribute space. Known data sets will be used to see if and how the time wave might discover known patterns and possibly reveal new temporal patterns as well, based on tasks executed by users. In relation to the task space (see Fig. 7.3) further work is ongoing in which existing temporal visualization methods are be analyzed on their strength and weakness in the context of the actual temporal problem of users. If successful this could lead to a kind of advisory system that assist the user to select a suitable graphic representation in time space dedicated to the user task one is dealing with.

Acknowledgements This research Supported by the National Natural Science Foundation for Young Scholars (No:41101445), and International Cooperation Project of Shaanxi Province Science Technology Department (No: 2011kw-08).

References

Aigner W, Miksch S et al (2007) Visualization time-oriented data – a systematic view. Comput Graph 31:401–409

Allen RB (1995) Interactive timelines as information system interfaces. Symposium Libraties, Japan

Andrienko G, Andrienko N (2006) Exploratory analysis of spatiall and temporal data. Springer, Berlin, Germany

Andrienko G, Andrienko N et al (2003a) Interactive maps for visual exploration of grid and vector geodata. ISPRS J Photogramm Remote Sens 57(5–6):380–389

Andrienko N, Andrienko G et al (2003b) Exploratory spatio-temporal visualization: an analytical review. J Vis Lang Comput 14(6):503–541

Bertin J (1983) Semiology of graphics: diagrams, networks, maps. University of Wisconsin Press, Madison

Brown IM (1998) A 3D interface for visualization of serial periodic data. ACM GIS'98, Washington DC USA

Carlis JV, Konstan JA (1998) Interactive visualization of serial periodic data. ACM UIST'98, San Francisco

Chi EH et al. (1998) Visualizing the evolution of web ecologies. ACM CHI'98, CA USA

Daassi C, Fauvet MC et al (2002) Multiple visual representation of temporal data. In Proceedings of the 13th international conference on database and expert systems applications, Springer-Verlag

Dykes J, MacEachren AM et al (eds) (2005) Exploring geovisualization. Elsevier Science & Technology Books, Amsterdam

Edsall R, Kraak MJ et al. (1997) Assessing the effectiveness of temporal legends in environmental visualization. GIS/LIS '97 Annual conference and exposition, Cincinnati, Ohio

Frank A (1994) Different types of "times" in GIS. GIS and Computational Science Perspectives

Goralwalla IA, Ozsu MT et al (1998) An object-oriented framework for temporal data models. In: Etzion O, Jajodia S, Sripada S et al (eds) Temporal database-research and practice, vol 1. Springer, Berlin, pp 1–35

Hägerstrand T (1967) Spatial process. University of Chicago, Chicago

Harris RL (1999) Information graphics: a comprehensive illustrated reference. Oxford University Press, New York

Harris S, Hetzler B et al. (2000) ThemeRiver: visualization theme changes over time. IEEE symposium on information visualization, USA, IEEE Computer Society

Harrison BL, Owen R, Baecker RM (1994) Timelines: an interactive system for the collection of visualization of temporal data. Graphics Interface' 1994, Canadian

Havre S, Hetzler B et al. (2000) ThemeRiver: visualizing theme changes over time. IEEE symposium on information visualization, Los Alamitos, USA, IEEE Computer Society

Koike Y, Sugiura A et al. (1997) TimeSlider: an interface to specify time point. In Proceedings of UIST '97

Li X, Kraak M-J (2008) The time wave. A new method of visual exploration of geo-data in time–space. Cartogr J 45(3):1–9

MacEachren AM (1995) How maps work: representation, visualization, and design. Guilford press, New York, USA

Mackinlay JD, Robertson GG et al. (1991) The perspective wall: detail and context smoothly integrated. ACM CHI'91, New York

Mackinlay J, Robertson G et al. (1994) Developing calendar visualizers for the information visualizer. In Proceedings of the 1990 ACM conference on computer-supported cooperative work, Los Angeles

Muller W, Schumann H (2003) Visualization methods for time-dependent data-overview. Winter simulation conference 2003, New Orleans USA

Peuquet DJ (2002) Representations of space and time. Guilford, New York

Plaisant C, Milash B et al. (1996) LifeLines: visualizing personal histories. Conference on human factors in computing systems, Vancouver, Canada ACM

Roberts JC (2005) Exploratory visualization with multiple linked views. In: Dykes J, MacEachren AM, Kraak MJ (eds) Exploring geovisualization, vol 8. Elsevier, London, pp 159–180

Roberts JC (2008) Coordinated multiple views for exploratory geovisualization. In: Dodge M, Mcderby M, Turner M (eds) Geographic visualization: concepts, tools and applications, vol 3. Wiley, Chichester, pp 25–48

Silva SF, Catarci T (2000) Visualization of linear time-oriented data: a survey. Web Information Systems Engineering, 2000

Tominski C, Abello J et al. (2003) Interactive poster: axes-based visualizations for time series data. Poster compendium of IEEE symposium on information visualization, IEEE

Wijk JV (2002) Image based flow visualization. SIGGRAPH, San Antonio, USA

Xiong R, Donath J (1999) PeopleGarden: creating data portraits for users. 12th annual ACM symposium on user interface software and technology (UIST '99), NC, USA

Chapter 8
Characterization and Detection of Building Patterns in Cartographic Data: Two Algorithms

Xiang Zhang, Tinghua Ai, and Jantien Stoter

Abstract Building patterns are important settlement structures in applications like automated generalization and spatial data mining. Previous investigations have focused on a few types of building patterns (e.g. collinear building alignments); while many other types are less discussed. In order to get better known of the building patterns available in geography, this paper studies existing topographic maps at large to medium scales, and proposes and discusses a comprehensive typology of building patterns, their distinctions and characteristics. The proposed typology includes linear alignments (i.e. collinear, curvilinear, align-along-road alignments) and nonlinear clusters (grid-like and unstructured patterns). We concentrate in this paper on two specific building structures: align-along-road alignment and unstructured clusters. Two graph-theoretic algorithms are presented to detect these two types of building patterns. The approach bases itself on auxiliary data structures such as Delaunay triangulation and minimum spanning trees for clustering; several rules are used to refine the clusters into specific building patterns. Finally, the proposed algorithms are tested against a real topographic dataset of the Netherlands, which shows the potential of the two algorithms.

X. Zhang (✉)
School of Resource and Environmental Sciences, Wuhan University, 129 LuoYo Road, Wuhan, People's Republic of China

Faculty of Geo-Information Science and Earth Observation, University of Twente, The Netherlands
e-mail: xzhang@itc.nl

T.H. Ai
School of Resource and Environmental Sciences, Wuhan University, 129 LuoYo Road, Wuhan, People's Republic of China
e-mail: tinghua_ai@tom.com

J. Stoter
Delft University of Technology, Delft, The Netherlands

Kadaster, Apeldoorn, The Netherlands
e-mail: j.e.stoter@tudelft.nl

A. Yeh et al. (eds.), *Advances in Spatial Data Handling and GIS*, Lecture Notes in Geoinformation and Cartography,
DOI 10.1007/978-3-642-25926-5_8, © Springer-Verlag Berlin Heidelberg 2012

8.1 Introduction

Building patterns are important features in urban and rural areas. The patterns can be used for people to infer the underlying geographic processes or the functional, socio-cultural and historic context of the settlement. Automated detection of visually significant building patterns is required for applications like automated map generalization, automated evaluation of generalized output, semantic enrichment of spatial databases, and spatial data mining. For example, collinear patterns have been extensively investigated (Boffet and Rocca Serra 2001; Christophe and Ruas 2002) to simplify and typify building groups. Higher-level semantics like terraced house can be automatically inferred based on detected building alignments and other criteria (Lüscher et al. 2009); but their detection approach is not widely applicable as the buildings in their case are adjacent, which is not usually the case.

As for the building pattern detection techniques, a remarkable investigation has been made by Zahn (1971) who explored the use of minimum spanning trees (MST) in recognizing point set patterns. After being successfully applied to some classical clustering problems for point sets, this technique was applied to detecting building clusters for map generalization (Regnauld 1996). Nevertheless, no interesting building pattern is recognized with this technique except for some general building clusters.

This paper firstly proposes a typology of building patterns by studying existing topographic maps (Sect. 8.2). Then Sect. 8.3 develops a graph-theoretic approach to the detection and characterization of two pattern types in the typology. The proposed algorithms are implemented and tested in Sect. 8.4. This paper ends with conclusions (Sect. 8.5).

8.2 Typology and Characteristics of Building Patterns

A typology of building patterns is needed as it formalizes our knowledge on settlement structures available in geography. The typology of the building patterns that can be observed in topographic maps and literatures are identified and characterized in Sect. 8.2.1, and then we focus specifically on align-along-road pattern, discussing its relationships to other alignments (Sect. 8.2.2).

8.2.1 Identifying and Characterizing the Typology

The building patterns discussed here are low-level, localized visual patterns, because they are important features that should be retained in map generalization. The typology (Fig. 8.1) is structured as follows. We define building patterns at the top level as building clusters of spatially proximate objects with similar geometrical (e.g. spacing, size, orientation and shape) and semantic properties. This extends the definition of point patterns in Zahn (1971). At the next level, we refine the building

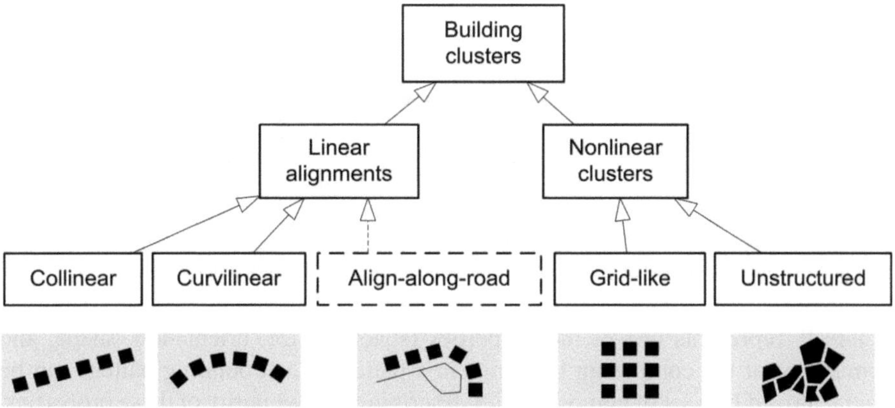

Fig. 8.1 Typology and schematic examples of building patterns

clusters into linear alignments and nonlinear clusters in terms of 'group shape'. In general, the linear alignments appear to be more elongated and the composing buildings are arranged on a path; while the nonlinear clusters appear to stretch themselves in two dimensions. At the finest level, the linear alignments are subdivided into collinear, curvilinear, and align-along-road patterns; the nonlinear clusters consist of grid-like and unstructured patterns.

We identify this typology from several sources. First, Gestalt principle of visual perception (Wertheimer 1923) is used to define building patterns intuitively (proximity, similarity, good continuity, etc. are the basic features). Second, by comparing with the typology proposed by others we found that this typology is more comprehensive and generic. As mentioned before, collinear alignment has been acknowledged in the literature. Anders (2006) also proposed a typology consisting of linear, circular, grid, star and irregular patterns. In our categories, his linear pattern is refined into collinear and curvilinear alignments; meanwhile, his circular pattern is a special case of a curvilinear alignment and the star pattern can be regarded as the combination of two crossing collinear alignments. Also, we propose a further type (i.e. align-along-road) which considers the relationship between alignments and the surrounding roads. The typology proposed in this paper is regarded to be generic also because it is based on the study of the maps (of the Netherlands, France, and Spain) in the EuroSDR on generalization project (Stoter et al. 2009a) and maps of China; a few examples are shown in Fig. 8.2. Other higher-level or global patterns can be seen as the combination of these low-level patterns.

This typology can be characterized as follows. According to previous work (Boffet and Rocca Serra 2001; Christophe and Ruas 2002; Ruas and Holzapfel 2003), the regularity of building clusters (termed homogeneity) can be realized using standard deviations of all the properties. In our approach, we implement the homogeneity by adopting the concept of coefficient of variance (CV):

$$Homogenity = STD(\mathrm{P}_i)/Mean(\mathrm{P}_i) \tag{8.1}$$

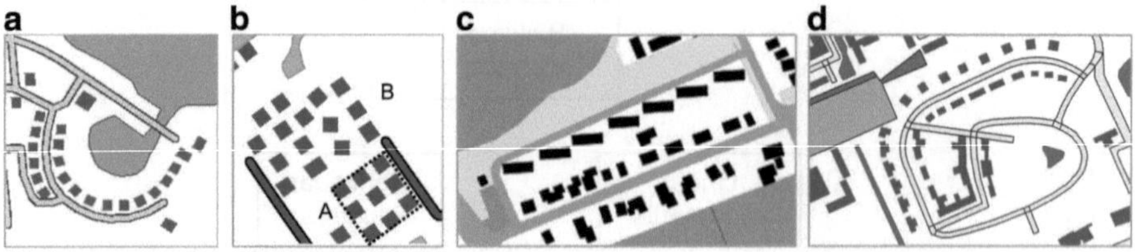

Fig. 8.2 Examples of building patterns on topographic maps: (**a**) curvilinear alignments, (**b**) grid and unstructured clusters, (**c**) collinear and (**d**) align-along-road alignments

where P_i represents one of the properties (spacing, size, orientation, shape, and semantics) of the composing buildings in a pattern. Each building pattern can be characterized by a set of homogeneities describing the regularity of these properties.

The use of CV applies to the properties of spacing, size, shape and semantics. This is not because we want to normalize the properties but rather because CV is a dimensionless number. It means that the homogeneity becomes a relative number invariant to the choice of measurement, which is also consistent with our perception about building patterns. For example, if two patterns have the same standard deviation of size, the one with larger mean size (i.e. smaller CV) is more homogeneous in terms of the size property. However, CV cannot be applied to orientation as it is meaningless to calculate CV for a cyclic variable. Therefore, the homogeneity of orientation is calculated using standard deviation.

The notion of homogeneity (8.1) is a common character shared by all pattern types. In addition, specific building patterns have their own characteristics. For collinear alignments, *straightness* (describing the sinuosity degrees of the paths) and *align angle* (describing the directions of the paths) are valuable. For curvilinear alignments, *smoothness* of the paths and their *curvature* descriptions should be emphasized. Align-along-road (AAR) patterns should know which roads they align along. Besides, it has another characteristic – the homogeneity of distances from the composing buildings to the aligned road. This property reflects the degree to which the AAR pattern is parallel to the road. For the two nonlinear clusters, unstructured clusters have no specific property, while grid patterns can be further characterized by *squareness* and *parallelism.* If we connect proximate buildings in a grid pattern with lines, we should get two sets of parallel lines, each intersecting with others approximately right-angled. Examples of the typology are illustrated in Fig. 8.2a–d.

8.2.2 Align-Along-Road Pattern and Its Relationship to Other Linear Alignments

As a result of studying existing topographic maps, we found that linear alignments are one of the most common features which are visually significant in urban and rural settlements [. . .]. We also found that the distinction between align-along-road patterns and collinear or curvilinear alignments is not always clear. On the one hand, buildings are usually located beside roads and streets, and therefore such collinear and curvilinear alignments (Fig. 8.2a,c) are also align-along-road patterns. On the other hand, collinear and curvilinear alignments may be viewed as part of align-along-road patterns, because normally align-along-road patterns are not regular in terms of curvatures. This means that an align-along-road pattern may be segmented into pieces of collinear and/or curvilinear alignments. Of course, there are also situations where collinear and curvilinear patterns are independent of roads.

In this paper, we limit ourselves to the detection and characterization of align-along-road and unstructured patterns. Among the remaining pattern types, some have been discussed by previous researchers while others should be addressed in future research.

8.3 Detection and Automatic Characterization Method

In this section, we propose two graph-theoretic algorithms for detecting align-along-road and unstructured patterns. Currently, semantics of buildings (e.g. relative importance) is not available in most topographic datasets (Stoter et al. 2009b) and is hence not included in this work. Nevertheless, our approach is able to include more information (e.g. semantics) for a more comprehensive detection and characterization framework.

This section first introduces the preliminary work which will be used by the subsequent detection and characterization (Sect. 8.3.1). Then the detection and automatic characterization for align-along-road (Sect. 8.3.2) and unstructured (Sect. 8.3.3) patterns are presented.

8.3.1 Basic Computational Tools

Refined Constrained Delaunay Triangulation (ReCDT). The ReCDT plays an important role in the subsequent detection and characterization process. We therefore introduce some fundamental computations based on this ReCDT structure in the first place. As shown in Fig. 8.3a, a CDT is constructed on buildings and roads, using their whole geometries (i.e. outlines) rather than centroids as constrained lines. Then the CDT is refined by inserting extra points to the constrained lines to avoid generation of long strip triangles; the interval between immediate inserted points is based on the minimal distance between all input points. Within a ReCDT, any line segment of any map object is a triangle edge; two map objects are considered as immediate neighbors only when they are connected by at least one triangle edge; the proximity relationship between two buildings and between buildings and roads are explicitly modeled by this structure (Fig. 8.3a).

Based on the proximity relationship between buildings, a proximity graph (Fig. 8.3d) is generated, based on which the Minimum Spanning Tree (MST) can be automatically derived (the next section). Note that in the proximity graph, as

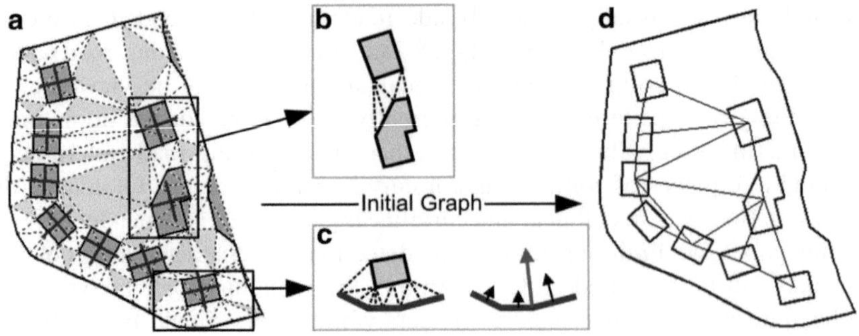

Fig. 8.3 (**a**) Constrained delaunay triangulation takes buildings (*grey polygons*) and roads (*bold lines*) as constrained objects; (**b**) distances defined on the incident triangles between proximate objects; (**c**) segments of a road along which the building aligns and an illustration of the normal direction of this part of the road (*the longest red arrow*); (**d**) the initial proximity graph from (**a**)

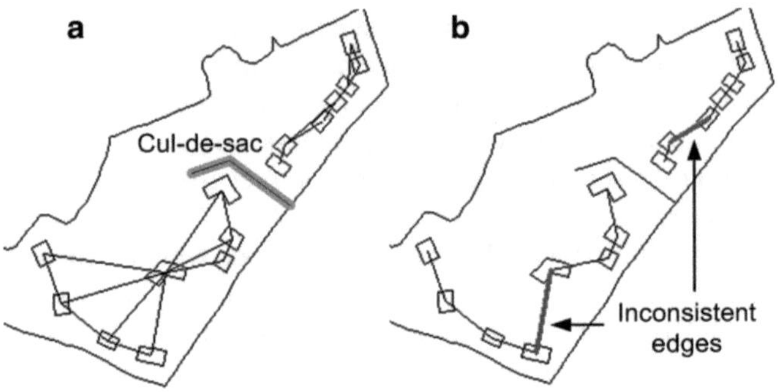

Fig. 8.4 Graphs derived from ReCDT: (**a**) proximity graph and (**b**) MST and inconsistent edges

well as in the derived MST, all edges have their weights which are based on the nearest distance between building outlines instead of building centroids, although for a graphic representation the graph vertices are placed at the centroids of buildings.

Several benefits can be drawn from this ReCDT structure. First, although being less efficient, the structure enables a better modeling of proximity relationship than constructing CDT using centroids of buildings, because ReCDT takes building forms into account. Second, because road geometries are also involved in ReCDT, the proximity graph is automatically segmented into sub-graphs so that no graph edge intersects the geometries of cul-de-sacs (Fig. 8.4a).

Theoretically, the overall computational efficiency for the pattern detection process (including constructing proximity graph, deriving MST, detecting and characterizing building patterns) can be compensated for by the ReCDT structure. This is because the calculation of the nearest distance between two proximate map objects can take advantage of the incident triangles in ReCDT (white triangles in Fig. 8.3a–c), enabling a more efficient computation. More specifically, for each incident triangle between two objects, a distance (d_i) can be easily computed; the nearest distance between the two objects is the minimum amongst all distances (Min($\{d_i\}$)). The time complexity of this calculation is O(t), where t is the number of incident triangles between two objects; while a trivial computation of the nearest distance between two buildings (as was used in Regnauld 1996) takes O(nm) time, where n and m are the numbers of the points of both buildings respectively. Since t is generally much smaller than nm that can be expected between two buildings (Fig. 8.3b) and especially between a building and a road (only a small part of the road is involved; see Fig. 8.3c), this actually greatly accelerates the computation of the nearest distance needed in the building pattern process.

Other calculations are needed to measure building orientations, the normal direction of the road a building aligned along and building shapes. First, the wall statistical weighting method (Duchêne et al. 2003) was implemented for computing building orientation; the authors have shown that this method is well suited for describing the orientations of typical manmade constructions (see the crosses in the buildings in Fig. 8.3a). Second, the normal direction of part of a road (the longest arrow in Fig. 8.3c) can be computed based on the addition of the normal vectors of its composing segments (the three shorter arrows in Fig. 8.3c) weighted by segment lengths. These two measures are crucial for the detection of align-along-road patterns. Finally, shape index (AGENT 1999) was implemented to characterize the complexity of building shapes.

Minimum Spanning Trees (MST). In graph theory, a spanning tree of an undirected graph G is a tree that contains all vertices of G; the weight of a spanning tree is defined as the sum of the weights of all its constituent edges. A MST of G is a spanning tree whose weight is the minimum among all spanning trees of G. Since a graph may not necessarily be connected (Fig. 8.4a), it has a union of minimum spanning trees for its connected components (Fig. 8.4b). The vertices of MST represent building features; the edges represent proximity relationships and are weighted by the nearest distances. According to Zahn (1971), Prim's algorithm (Prim 1957) is more efficient than Kruskal's algorithm, so the former was implemented here to derive MST from proximity graph.

A fundamental concept in MST for out pattern recognition process is the so-called *inconsistent edge* proposed by Zahn (1971), where he has shown evidence that MST and the inconsistent edge are perceptually significant in point set clustering. Intuitively, an inconsistent edge is an edge of MST whose weight is significantly larger than the mean weight of nearby edges on both sides of this edge (e.g. bold edges in Fig. 8.4). Formally, it can be defined as follows:

$$edge_i = \begin{cases} \text{inconsistent}, & \text{if } w_i > I_l \cap w_i > I_r \\ \text{consistent}, & \text{otherwise} \end{cases} \tag{8.2}$$

where I is a measure that can be defined on both left (I_l) and right (I_r) sides of $edge_i$:

$$I = \max(f \times mean_{\text{weight}}, mean_{\text{weight}} + n \times std_{\text{weight}}) \tag{8.3}$$

An edge is said to be inconsistent if its weight exceeds the mean weight ($mean_{weight}$) of its neighboring edges (within p depth) on both sides by n units of standard deviation (std_{weight}) and further more if it is at least f times as large as the weights on both sides (Zahn 1971, p. 82). Given a normal distribution, there is less than 1% chance that an edge weight would exceed $mean_{weight}$ by 3 unit of std_{weight}. Therefore, $n \geq 3$ is regarded to be statistically significant for defining I. In some cases where std_{weight} approaches to zero, the factor $f > 1$ ensures that the inconsistent edge is still significant with respect to $mean_{weight}$. A detailed discussion on the parameterization issue is available in Zahn (1971). By pruning the inconsistent edges, clusters emerge naturally.

However, the inconsistent edge concept alone is not sufficient for detecting meaningful building patterns. We did pilot experiments and found that no matter how to adjust and combine the parameters, most of the building patterns do not automatically show themselves up by simply pruning the inconsistent edges. The reason is mainly that the MST-based techniques, which only enforce the proximity principle (Zahn 1971), do not always lead to a result that resembles human perception. Therefore other principles of perceptual organization (Wertheimer 1923) must also be integrated for further processing depending on what kinds of building patterns are to be recognized.

The parameterization of inconsistent edge appeared to be not critical in recognizing building patterns. To make sure that most significant edges are identified and pruned while less significant ones are left out for further decisions, a conservative set of parameters ($p = 2$, $n = 3$, and $f = 2.0$) was finally determined. Although other values are also possible, our experiments show that the variations make little difference for the final detection results, probably because the difference made in the pruning step can be made up for by the subsequent processes (see Sect. 8.3.2).

8.3.2 Align-Along-Road Building Pattern

There is a generic graph-theoretic framework for the detection of all linear alignments in our work, that is, the detection is achieved by tracing paths in MST, and the trace of the patterns should satisfy different constraints that are appropriate for each type of the patterns. A path can be formally defined as a sequence of vertices in a MST (e.g. the line walking through the buildings on the bottom of Fig. 8.5a); and the path angle at vertex v_i is defined by the angle between $vector(v_{i-1}, v_i)$ and $vector(v_i, v_{i+1})$ in that path (Fig. 8.5a).

Detection of the align-along-road pattern. The basic idea of detecting align-along-road patterns is that a path should be traced from the pruned MST and the buildings on the path should be close enough to a nearby road. Therefore, we formulate the detection of the align-along-road pattern as a path tracing procedure that advocates the same aligned road, proximity, size similarity and

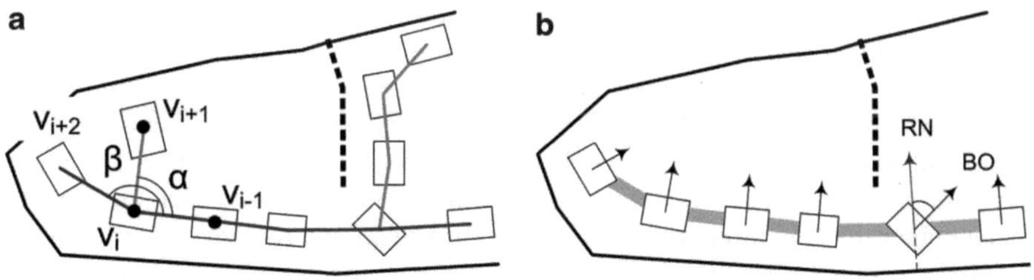

Fig. 8.5 Illustrations of the detection (**a**) and the characterization (**b**) of align-along-road pattern

path angle constraints (Algorithm 1). The choice of the constraints is justified as follows.

Algorithm 1: Detecting the align-along-road pattern

```
Input: buildings; partition roads; constraints
Output: Collection of align-along-road patterns (AARP)
1: calculate CDT and then MST for the input data;
2: prune the inconsistent edges from the MST;
3: trace paths in the pruned MST as follows:
foreach v0 = vertex of degree 1 or vertex of degree 3 in MST do
    loop "initialize AARP"
        v1 = adjacent vertex of v0;
        if v0,v1 align along the same closer road then
            initialize AARP with edge(v0, v1);
            break;
        else
            v0 = v1;
    endloop;
    loop "add new buildings to AARP"
        if v1 is of degree 1 then
            add AARP to Collection if AARP contains more
            than 3 buildings;
            break;
        if v1 is of degree 3 then
            select a vertex v2 with which the edge(v1, v2)
            forms larger path angle with its predecessor edge.
        else
            v2 = adjacent vertex of v1
        check edge(v1, v2) w.r.t. the constraints for AARP;
        if edge(v1, v2) exists then
            add edge(v1, v2) to AARP;
            v0 = v1; v1 = v2; v2 = NULL;
        else
            add AARP to collection if AARP contains more
            than 3 buildings;
            empty AARP; renew AARP with edge(v1, v2) if
            the two vertices share a road;
            v0 = v1; v1 = v2; v2 = NULL;
    endloop;
4: combine the detected AARP if possible;
```

In general, we found that shape and orientation are not as dominant as aligned road, proximity, and size similarity in determining a collinear group. The first constraint is that buildings align along the same road can be recognized as an align-along road pattern. The second constraint is that buildings composing the alignment should have similar sizes. Shape similarity is discarded as most buildings have similar man-made forms, and those with complex forms, which are usually much bigger than the ordinary buildings, can be filtered out by size similarity in the first place. On the other hand, the measure that captures the generic nature of the shape in map generalization is currently not available. The orientation constraint is also dropped, since the theoretically maximum deviation between buildings computed by the wall statistical weighting method is 45° (Duchêne et al. 2003), which has much less impact on the perception of align-along-road patterns than the proximity constraint. Note however that, the shape and orientation properties are integrated in the characterization of the patterns.

The selected constraints work together for the pattern detection in the following way. In general, the constraints work on each new edge occurring in the tracing process (Algorithm 1), determining if the new edge can be added to an existing alignment. To begin with, aligned road is checked using the information stored in ReCDT; if buildings connected by the new edge share the same road, the tracing procedure proceeds to the next edge; otherwise the tracing stops. Then, the proximity constraint ensures that if the new edge is inconsistent with respect to the edges in the existing alignment, the edge should not be added to this alignment. Finally, the size similarity constraint ensures that the size of the two buildings connected by the new edge should not be too large (i.e. bigger building/smaller building < *size_contrast*). It is worth mentioning that if in the iteration there are more than one candidate edges available (e.g. $edge(v_i, v_{i+1})$ and $edge(v_i, v_{i+2})$ in Fig. 8.5a), an edge with larger path angle (α, $\beta \in [0°,180°]$; $\beta > \alpha$) should be added to the existing alignment in order to keep the good continuity principle (Wertheimer 1923).

In the step 4 of Algorithm1, if two detected alignments approach to each other at their ends a combination is recommended. This combination can be done by introducing an edge connecting the proximate ends of both alignments, only if on the one hand the two groups align along the same road and on the other the distance of the introduced connecting edge is consistent with respect to both alignments. The parameter showing promising results for align-along-road pattern is *size_contrast* = 3.2.

Characterization of the align-along-road pattern. The characterization of align-along-road building patterns is by applying (8.1) to properties such as spacing, size, shape, and distance to the aligned road. For homogeneities, mean values are firstly calculated for these properties and then the homogeneitis are computed. Note that both spacing (i.e. distance between buildings) and distance to the aligned road are computed using the nearest distance calculation described in Sect. 8.3.1.

The calculation of the homogeneity of orientation is described as follows. The orientation is considered to be more regular if the buildings change their orientations right according to the normal directions (e.g. RN in Fig. 8.5b) of the local road segments that they align along. For each building in the pattern, an angle deviation ∈ [0°,45°] is calculated between RN (Normal of Road segments) and BO (Building Orientation); the homogeneity of orientation is then computed by the standard deviation of all the deviations.

8.3.3 *Unstructured Clusters*

Detection of unstructured clusters. Unstructured clusters are also a common feature on topographic maps, especially at larger scales (1:10 k–1:50 k).

Algorithm 2: Detecting unstructured clusters

Input: buildings; partition roads [optional]; *postconditions*
Output: *Collection* of unstructured clusters (UC)

1: calculate CDT and then MST for the input data;
2: prune the inconsistent edges from the MST;
3: prune the edges connecting buildings whose size difference $> size\_contrast$; if $FragRatio > 0.5$, the edges should not be pruned;
4: organize connected subgraphs and populate UC with them;
5: filter out those UC that cannot pass *postconditions*, and add remaining UC to *Collection*;

Algorithm 2 detects unstructured clusters and is realized through pruning edges that are inconsistent and edges that connect two buildings whose size difference exceeds *size_contrast*. The resulting graphs, however, are just candidates that have to be further tested against several *postconditions*. The first postcondition is the number of buildings contained. In our experiments, we define that at least three buildings form a cluster or pattern, as only in this case the calculation of mean and standard deviation required by the detection and characterization is meaningful. A second postcondition is black-and-white ratio, which is defined as follows:

$$BWRatio = \frac{\sum \text{Area}(b_i)}{\text{Area}(\text{ConvexHull}(UC))} \tag{8.4}$$

where b_i∈UC (Unstructured Clusters). This postcondition protects some wriggling linear alignments from being recognized as unstructured clusters (Fig. 8.6a). The third postcondition is termed fragmentation ratio (FragRatio), which is the ratio between the number of pruned edges and the number of total edges in an initial cluster. This postcondition should be applied in step 3 (Algorithms 2) to prevent clusters from being over-fragmented, as in Fig. 8.6b that small and big buildings are alternately arranged.

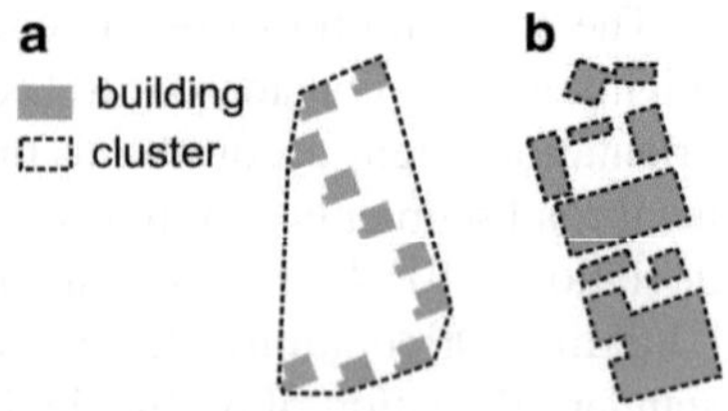

Fig. 8.6 Examples of incorrectly detected unstructured clusters

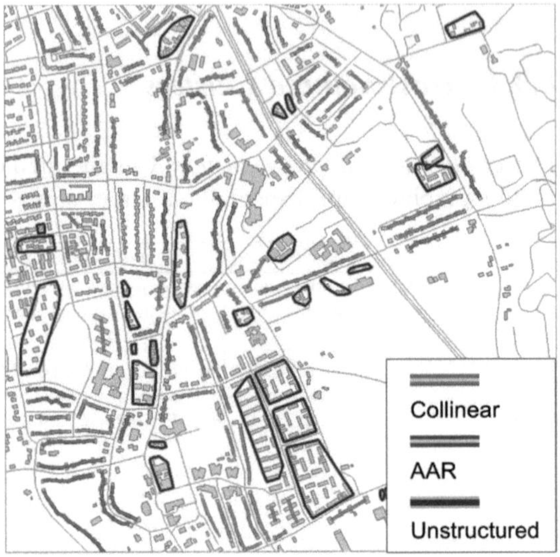

Fig. 8.7 Test case and detected collinear, align-along-road, and unstructured patterns

Characterization of unstructured clusters. Despite the homogeneity of orientation which is calculated based on absolute orientations of buildings, the homogeneities of spacing, sizes and shapes are exactly the same to those calculated for align-along-road patterns. The algorithms for the detection and characterization were implemented and results are presented in Sect. 8.4.

8.4 Implementation and Results

We implemented the two proposed algorithms in an interactive generalization prototype system using C++. In addition, an algorithm to detect collinear building patterns was also implemented. The test case and detection results for collinear, align-along-road (AAR), and unstructured patterns are shown in Fig. 8.7. There are 151 patterns detected from the test dataset, where there are 94collinear alignments, 34 unstructured clusters, and 23 align-along-road patterns. It is noticeable that some building groups can be viewed as both AAR and collinear alignments.

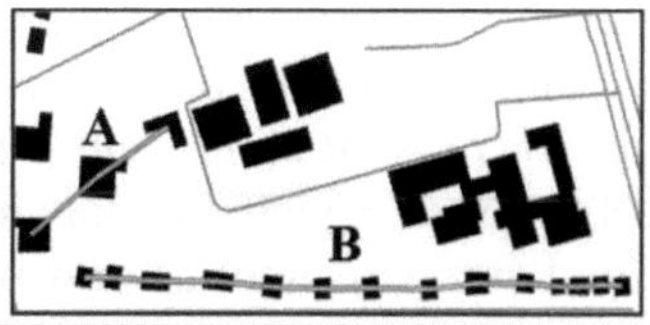

id	spacing	size	orientation	shape
A	0.06	0.32	9.74°	0.43
B	0.52	0.26	5.57°	0.11

Fig. 8.8 Measured characteristics of collinear alignments

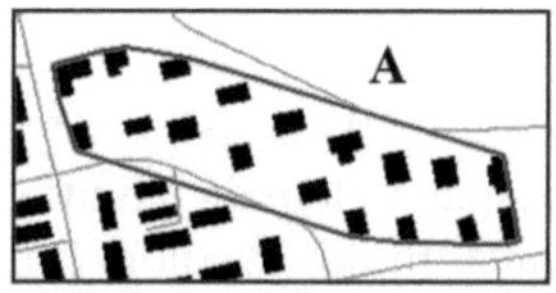

id	spacing	size	orientation	shape
A	0. 22	0. 22	2. 03°	0. 24

Fig. 8.9 Measured characteristics of an unstructured cluster

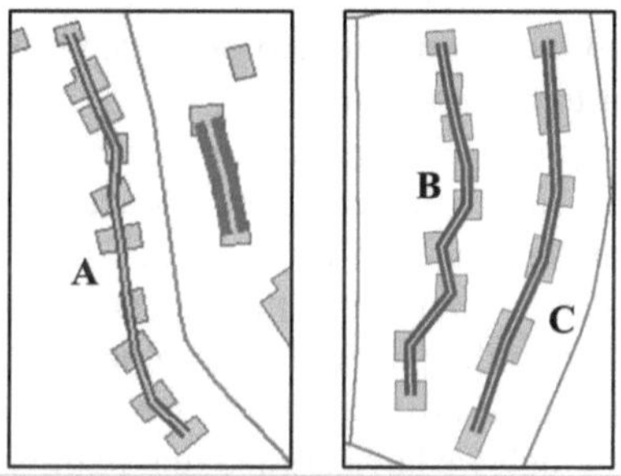

id	spacing	dis2road	size	ori1	ori2	shape
A	0. 39	0. 13	0. 28	4. 34°	9. 94°	0. 15
B	0. 39	0. 78	0. 15	2. 4°	2. 36°	0. 09
C	0. 07	0. 11	0. 28	2. 14°	12. 95°	0. 18

Fig. 8.10 Measured characteristics of align-along-road patterns

The characterization results are shown in Figs. 8.8, 8.9, and 8.10. A general observation is that the measured values for spacing, size, shape, and dis2road (values∈[0,1]) and the values for orientation (values ∈ [0°,45°]) are consistent with our perception of these detected patterns. For examples, the measured homogeneities of size, orientation, and shape for the collinear patterns in Fig. 8.8 confirms that alignment B is more visually significant than A in most aspects; the building cluster shown in Fig. 8.9 is confirmed by the measured homogeneities of this unstructured cluster, as all the homogeneity values are relatively low (below a quarter).

For the align-along-road patterns as shown in Fig. 8.10, several observations can be made. First, 'dis2road' column confirms that this characteristic is more homogeneous for A and C than B, as B is aligned along the road on the left. Second, an extra orientation ('ori2') is calculated based on absolute values, in order for the readers to compare it with the orientation that changes with aligned roads ('ori1'). The result shows that 'ori1' is a better indicator than 'ori2' for characterizing the orientation of AAR alignments. The result of 'ori1' indicates that pattern A and C in Fig. 8.10 are acceptable alignments, which agrees with our perception. This conclusion however cannot be drawn from 'ori2'.

8.5 Discussion and Conclusion

This paper firstly proposes a typology of building patterns available in cartographic and geospatial domain based on the study of existing topographic maps and related literatures. In this typology, fundamental visual patterns have been identified and their characteristics been discussed. This is an important step towards formalizing our knowledge on the building patterns in this field. Secondly, two graph-theoretic algorithms are proposed to detect two pattern types in the typology, namely the align-along-road and unstructured patterns. The detection and characterization method was implemented and tested using real topographic datasets. The results appear to be promising.

The proposed approach is generic in the sense that visually important building patterns can be detected and characterized no matter the buildings are topologically adjacent or not, even if the spatial objects are represented by points, since all these cases can be handled by the ReCDT structure. This approach can be further applied to detecting the patterns of archipelago, lakes and ponds, by adjusting (in part) the constraints or postconditions used in the algorithms.

Although we have argued that in most cases collinear and curvilinear alignments can be viewed as align-along-road patterns, the detection of align-along-road patterns depends on the availability of road data. Besides, the characterization of align-along-road patterns is still too general. That is, important characteristics, like the straightness and main angle of collinear alignments, and the smoothness and curvature descriptions of curvilinear alignments, would be meaningless if those linear patterns were recognized as align-along-road patterns. Therefore, future work will be focus on the detection and characterization of collinear and curvilinear alignments which are the most common feature that can be found in urban and rural areas on topographic maps.

Another important issue that was not addressed is that, multiple algorithms will definitely result in detection results that overlap and conflict each other. For example, a group of buildings may at the same time be recognized as collinear, curvilinear and align-along-road alignment by different specific algorithms. The results from multiple algorithms should be combined and conflicts removed before the pattern detection approach can be practically useful. An important application is automated evaluation of building pattern preservation (see Zhang et al. 2010) in order to assess the generalization of settlement structures (e.g. building alignments).

Acknowledgements The work was partly supported by the 863 Program (Grant No. 2009AA121404). Faculty of Geo-Information Science and Earth Observation (ITC) of the University of Twente, which funds the first author as a PhD student, is gratefully acknowledged.

References

AGENT (1999) Selection of basic measures, deliverablec1. Technical Report

Anders KH (2006) Grid typification. In: Riedl A, Kainz W, Elmes GA (eds) Progress in spatial data handling, Springer, Vienna, pp 633–642

Boffet A, Rocca Serra S (2001) Identification of spatial structures within urban block for town qualification. In: ICC, vol 3, Beijing, pp 1974–1983

Christophe S, Ruas A (2002) Detecting building alignments for generalisation purposes. In: Richardson DE, van Oosterom P (eds) Advances in spatial data handling (SDH 2002). Springer, Ottawa, pp 419–432

Duchêne C, Bard S, Barillot X, Ruas A, Trévisan J, Holzapfel F (2003) Quantitative and qualitative description of building orientation. In: 5th workshop on progress in automated map generalization, Pairs

Lüscher P, Weibel R, Burghardt D (2009) Integrating ontological modelling and bayesian inference for pattern classification in topographic vector data. Comput Environ Urban Syst 33 (5):363–374

Prim RC (1957) Shortest connection networks and some generalizations. Bell Syst Tech J 36:1389–1401

Regnauld N (1996) Recognition of building clusters for generalization. In: Kraak MJ, Molenaar M (eds) Advances in GIS research II (Proceedings of 6th SDH'96, Delft), Taylor & Francis, London, pp 4B.1–4B.14

Ruas A, Holzapfel F (2003) Automatic characterisation of building alignments by means of expert knowledge. In: ICC, Durban, pp 1604–1515

Stoter J, Burghardt D, Duchene C, Baella B, Bakker N, Blok C, Pla M, Regnauld N, Touya G, Schmid S (2009a) Methodology for evaluating automated map generalization in commercial software. Comput Environ Urban Syst 33(5):311–324

Stoter J, van Smaalen J, Bakker N, Hardy P (2009b) Specifying map requirements for automated generalization of topographic data. Cartogr J 46(3):214–227

Wertheimer M (1923) Laws of organization in perceptual forms. In: Ellis WD (ed) A source book of gestalt psychology. Routledge & Kegan Paul, London, pp 71–88

Zahn CT (1971) Graph-theoretical methods for detecting and describing gestalt clusters. Comput IEEE Trans C20(1):68–86

Zhang X, Stoter J, Ai T, Kraak MJ (2010) Formalization and data enrichment for automated evaluation of building pattern preservation. In: Joint international conference on theory, data handling and modelling in geoSpatial information science, Hong Kong, vol XXXVIII, Part 2, pp 267–272

Chapter 9
Formalization and Data Enrichment for Automated Evaluation of Building Pattern Preservation

Xiang Zhang, Jantien Stoter, Tinghua Ai, and Menno-Jan Kraak

Abstract Automated evaluation of generalization output relies largely on well defined map specifications or cartographic constraints to formalize quality criteria or user requirements, and on enriched information to fully automate the process. Previous studies suggest that the formalization and evaluation of legibility constraints (i.e. those improving the readability of maps) are relatively easier than preservation constraints (i.e. those preserving important geographic characteristics and patterns). Patterns are important structures and should be taken into account by the evaluation process. This paper aims at a methodology in which the preservation of building patterns can be evaluated automatically. Three major difficulties in the process are identified and addressed: (1) pattern classification, characterization and detection, (2) pattern matching, and (3) constraint formalization. In addition, the

X. Zhang (✉)
School of Resource and Environmental Sciences, Wuhan University, 129 LuoYo Road, Wuhan, People's Republic of China

Faculty of Geo-Information Science and Earth Observation, University of Twente, Enschede, The Netherlands
e-mail: xzhang@itc.nl

J. Stoter
Delft University of Technology, Delft, The Netherlands

Kadaster, Apeldoorn, The Netherlands
e-mail: j.e.stoter@tudelft.nl

T.H. Ai
School of Resource and Environmental Sciences, Wuhan University, 129 LuoYo Road, Wuhan, People's Republic of China
e-mail: tinghua_ai@tom.com

M.-J. Kraak
Faculty of Geo-Information Science and Earth Observation, University of Twente, Enschede, The Netherlands
e-mail: kraak@itc.nl

A. Yeh et al. (eds.), *Advances in Spatial Data Handling and GIS*, Lecture Notes in Geoinformation and Cartography,
DOI 10.1007/978-3-642-25926-5_9,

knowledge of transition events describing allowed changes to building patterns at scale transitions is obtained via studying existing topographic maps (from 1:10 k to 1:100 k). Based on this knowledge, automated matching of corresponding building patterns is done much easier. Finally, the proposed methodology is implemented and validated by being applied to the evaluation of an interactively generalized dataset against its initial dataset. The results show the potential of the methodology and that the generalized patterns tend to remain the same or become more regular with respect to the initial ones. We also identify further improvement for practical use in an overall evaluation process to indicate acceptable generalization solutions.

9.1 Introduction

In map generalization smaller scale maps are derived from larger scale maps. This intelligent information management process involves a combination of data reduction and simplification related techniques in order to suppress unnecessary detail. At the same time geographic patterns and structures are emphasized to achieve a clear view of information that resembles the characteristics in reality as much as possible. Because settlement structures (groups of buildings exhibiting special arrangements that may imply functional or socio-cultural aspects of the original design of the settlement) are significant for the topographic maps ranging from large to medium scales, preservation of building patterns is an important map specification (i.e. cartographic constraint) in the generalization process. The constraint aims at keeping important settlement structures such as clusters and alignments (e.g. buildings aligned along roads and buildings in rows) and representing them on generalized maps.

Automated evaluation of generalization output in general aims to assess whether or to what extent the output satisfies the cartographic constraints according to automatically measured indicators (e.g. size and shape). Automated evaluation relies firstly on constraints formalization. From related work (Weibel and Dutton 1998; Burghardt et al. 2007; Stoter et al. 2009a) we can conclude that while patterns are routinely considered in conventional cartography, the relevant constraints (e.g. on networks, patterns, and distributions of map features) are harder to formalize and evaluate automatically than legibility constraints (e.g. minimum dimension of an object required to distinguish it on the map). This paper therefore concentrates on automated evaluation of building pattern preservation, whereby the formalization and data enrichment techniques that enables the evaluation are explored. To apply the evaluation, interactively generalized building patterns are evaluated against initial patterns to see how the generalized patterns meet the specification.

Difficulties in automated evaluation of building pattern preservation are manifold. First, the written specifications concerning building patterns that intend for interactive generalization are not easy to formalize, as they are meaningful for cartographers but not for computers. A constraint saying for example that building alignments should be preserved can be interpreted and then generalization operations applied by cartographers based on their knowledge and/or the generalization situation. Formal knowledge for computers to characterize the patterns and to describe their changes at scale transitions is necessary but not yet available for automated evaluation. Therefore, the written specifications need to be enriched for automated processes.

Second, building patterns are not stored as database objects in topographic datasets. Consequently to automatically evaluate generalized building patterns, the datasets have to be enriched with building pattern instances recognized from building groups by either pattern recognition techniques or visual inspection. Previously, Regnauld (1996) detected building clusters for building typification. However, except for general clusters, no refinement and characterization were made. Lüscher et al. (2009) developed a promising approach to detect higher level semantics (e.g. terraced house) based on building alignments and other criteria. They detected the alignments by assuming that buildings in the alignments are topological adjacent, which is however not usually the case (as those discussed in this paper) and is hence not widely applicable. Besides, the knowledge on a comprehensive view of building pattern types is still not clear.

A third difficulty in automated evaluation of building pattern preservation is the lack of linkages between corresponding patterns at different scales. In the context of the evaluation of generalization output (Mackaness and Ruas 2007), the links are especially necessary for the automated evaluation of preservation constraints (see also Stoter et al. 2009a) to give the system the knowledge as to which pattern instances at different scales correspond to one another. Currently, topographic databases are maintained at many national mapping agencies without explicit maintenance of links between map features. Additional effort is hence needed to establish the links.

The links can be created via generalization operators as part of the generalization process (during the process it is still known which representations are generalized into which representations at smaller scales) and via data matching (Hampe et al. 2003). An example of linking objects during generalization is that as part of aggregation a link can be established between the aggregated object and the initial objects. Most of such links, however, are currently meant for temporary use and are not stored as final output. In addition, links as a result of the generalization is highly complex, where different operators are triggered in an iterative and/or interactive manner, and thus it is almost impossible to maintain those links during the process. Besides, pattern objects may not always be explicitly handled by the operators, and hence the links between them are less likely to be maintained automatically. Consequently, to be able to automatically evaluate preservation of building patterns, data matching techniques are required for creating links between corresponding pattern objects.

The remainder of this paper is organized as follows. First, this paper presents our methodology that realizes two important above-mentioned prerequisites for automated evaluation of building pattern preservation in generalization (the implicit building patterns and their characteristics are made explicit, which was already proposed and tested in Zhang et al. (2010) and is summarized in Sect. 9.2); the two issues concern the links and the formalization of the constraint, which are addressed in Sect. 9.3 on the methodology for identifying, formalizing and evaluating building pattern preservation. Then, the methodology are implemented and validated by applying it to interactively generalized maps, where building patterns are assumed to have been well preserved (Sect. 9.4). This paper ends with discussion and conclusions (Sect. 9.5).

9.2 Classification and Characterization of Building Patterns

This section presents the formalization of building patterns that was realized by classifying building patterns and characterizing them. To better understand the concept of building patterns in generalization, we studied the map specifications written in the Dutch Topographic Survey (part of the Netherlands' Kadaster) and the interactively generalized maps (including those of the Netherlands, France, Spain, and China), as well as cookbooks on map generalization and previous studies on building pattern types (e.g. Anders 2006). The result shows that not only alignments but also other forms of building groups are cartographically significant and should be kept during generalization. Based on the findings we proposed a typology of building patterns which is discussed in greater detail in Zhang et al. (2010), where algorithms to detect them are also presented. The typology is important to the formalization of building pattern preservation constraint, as it enriches the patterns type with important, formal information. This information helps to accommodate different building patterns in pattern matching and evaluation. A UML class diagram of the proposed typology is shown in Fig. 9.1.

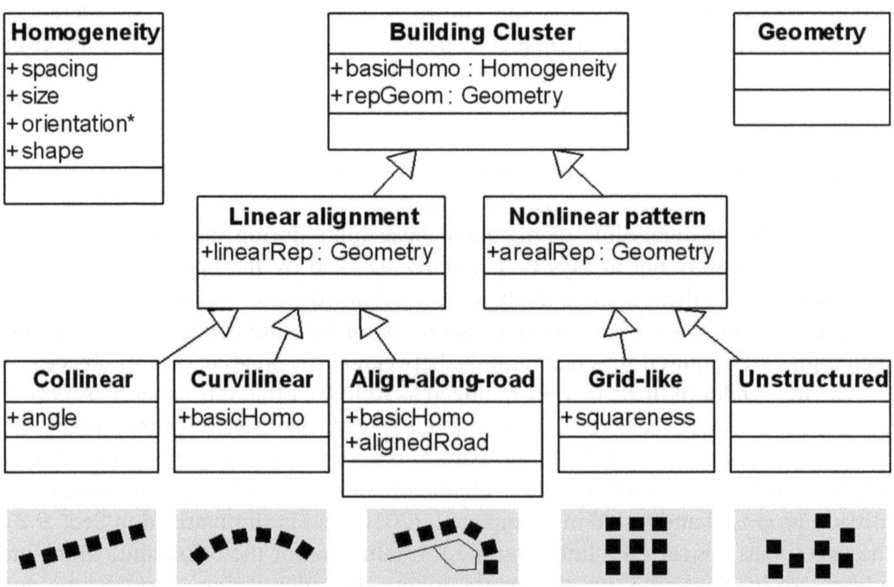

Fig. 9.1 A UML model of the proposed pattern classification and examples of each pattern type

The UML diagram models the patterns of the typology with their representational geometries and characteristics (which can be automatically detected by the algorithms described in Zhang et al. (2010)). We distinguish between five types of patterns: three kinds of linear patterns (collinear, curvilinear, and align-along-road alignments) and two kinds of nonlinear patterns (grid-like patterns and unstructured clusters). The properties that all these types have in common are representational geometries (denoted as 'repGeom') and the notion of homogeneity, reflecting the regularity of the patterns (denoted as 'basicHomo' in the UML diagram). Two types of representational geometries are used:

- Linear representations (denoted as 'linearRep'): Skeletons of linear alignments; the skeletons are derived from the minimum spanning trees of building centers (see Zhang et al. 2010).
- Areal representations (denoted as 'arealRep'): Convex hulls of nonlinear patterns; buffers of the above-mentioned skeletons in cases where linear patterns are to be matched with nonlinear ones (see Sect. 9.3.3).

The common attribute 'basicHomo' consists of homogeneities of spacing (between proximate buildings), sizes, orientations, and shapes. The homogeneity can be calculated by standard deviation (Ruas and Holzapfel 2003). This also means that when a pattern becomes more regular its homogeneity decreases, and vice versa. Note however that the homogeneity of orientations should be implemented differently for curvilinear and align-along-road patterns. This is why the two classes override their parent class's 'basicHomo' property. For example, the orientation is homogeneous for a collinear pattern if all the composing buildings have the same orientation; while a curvilinear pattern is regarded to be homogeneous only if the orientations of buildings vary right according to the 'path' formed by the buildings (Fig. 9.2(2)). Besides, specific characteristics in the diagram are explained as follows: the 'angle' of a collinear alignment is the main direction of its skeleton; any align-along-road pattern should be attached to an 'alignedRoad'; the 'squareness' quantifies the degree to which buildings align in a rectangular grid layout.

Note that, these characteristics are not unique properties to distinguish different pattern types (e.g. 'basicHomo' is shared by all types of patterns). Instead, differences between pattern types are specified in the detection algorithms; the characteristics are measured after the detection process to be able to compare and evaluate the patterns.

The UML diagram formalizes different pattern types and makes a step further towards automated evaluation of building patterns preservation, since the type information of the patterns are now explicitly available to the system, thus enabling the system to assess whether the change from one pattern type to another is acceptable or not (Sect. 9.3).

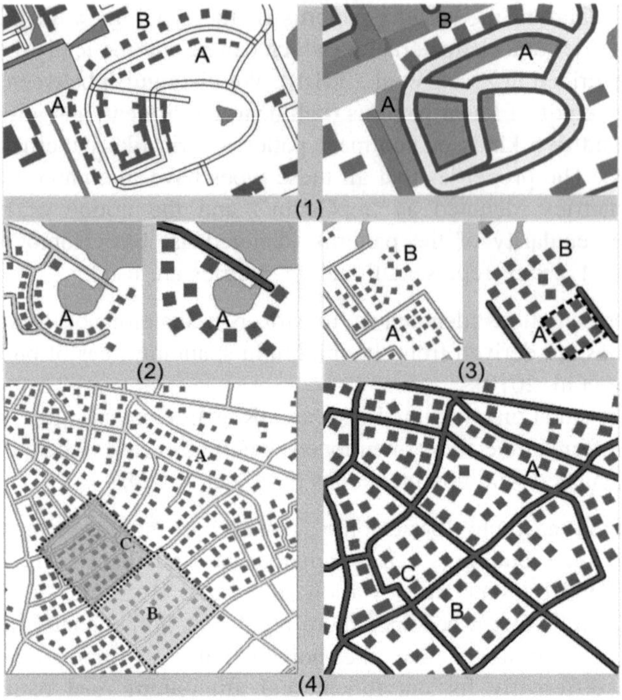

Fig. 9.2 The cases illustrating typical changes of building patterns at scale transitions from (**1**) TOP10NL to TOP50NL and (**2**)–(**4**) TOP5ONL to TOP100NL (examples from maps of the Netherlands' Kadaster)

9.3 Methodology for Identifying, Formalizing and Evaluating Building Pattern Preservation

After the building patterns are detected and characterized, the next step is to assess whether the generalized map preserves the regularity (homogeneity properties) of the initial patterns in an acceptable way. This requires firstly the knowledge on how the patterns change at scale transitions and hence which patterns at different scales are linked (Sect. 9.3.1, 9.3.2 and 9.3.3), and secondly the methods to (a) automatically evaluate the constraints and (b) to interpret the quantitative evaluation results (Sect. 9.3.4).

Note that the first issue is addressed through data matching of building pattern instances which is different from data matching of individual buildings as it links two building groups rather than individual buildings at different scales together. In our approach correspondence relationships (i.e. multiplicity in Bobzien et al. 2008) are modeled at the group level. For example, if an unstructured cluster containing more buildings is generalized into another cluster containing fewer buildings then a 1:1 relationship is modeled between the clusters rather than between the composing buildings. A further issue is that one type of pattern may become another type after the generalization while still being acceptable. For instance, part of an unstructured cluster may become a collinear alignment (i.e. become more regular). The change of pattern type brings about more difficulties to the matching process. To address the issue, Sect. 9.3.1 first analyzes interactively generalized maps (where the patterns are assumed to having been generalized in an acceptable way) to identify possible changes of building patterns at scale transitions (i.e. diminished, transformed, or preserved). From closer observations, Sect. 9.3.2 further formalizes the allowed changes of the preserved building patterns at scale transitions (transition events), which helps to match potential pattern candidates with different types (Sect. 9.3.3).

9.3.1 Identification and Formalization of Changes of Building Patterns During Generalization

Topographic maps were analyzed for the following reasons: (1) to understand why and how building patterns change at scale transitions; (2) to enrich the available written specifications for generalizing building patterns. In this analysis we visually detect building clusters and alignments in both initial and generalized data and then qualify several changes of the patterns at scale transitions: do they diminish, are they preserved or are they transformed?

To answer these questions two scale transitions (1:10–1:50 k and 1:50–1:100 k) in the topographic maps of the Netherlands were studied. Maps at scale 1:10 k (supported with an object oriented database, called TOP10NL*)* is the most detailed topographic database, from which maps at scale 1:50 k (supported with TOP50NL database) are interactively generalized. Maps at scale 1:100 k (supported with TOP100NL database) are interactively generalized from 1:50 k maps. By comparing building patterns at these scales, the following transitional knowledge was obtained (due to limited space only a few examples are shown in Fig. 9.2).

At a top level, we observe three kinds of pattern transitions. First, some building patterns are diminished. For example, a collinear pattern consisting of three buildings is diminished after two buildings have been removed. The reasons for this may be due to other important constraints, e.g. respecting the minimum distance constraint may remove the buildings that are too close. Second, some building patterns are transformed into built-up areas. This is common in the Dutch case due to a constraint stating that areas of high building densities should be replaced by built-up areas (Stoter et al. 2009b). We also found that some built-up areas are based on the areas that are not fully surrounded by the road network but are based on extents covered by building alignments in initial with higher densities (1A in Fig. 9.2).

The third important observation is that, the generalization (e.g. simplification, typification, removal, aggregation, etc.) of buildings composing a building pattern can cause changes to the pattern. In our study we focus specifically on tolerated changes to building patterns as a result of these changes. That is that the patterns being diminished or transformed into built-up areas are regarded as patterns not being kept and hence they will not be discussed in the following sections. Section 3.2 (formalization of transition events) further explores why and how the patterns are preserved, or in other words, which modifications are allowed to preserve the pattern.

9.3.2 Formalizing Observed Transition Events of Pattern Preservation

From closer observations of the above visual analysis of the map series, a set of transition events describing allowed changes for building pattern preservation are formalized in Table 9.1. Representative cases of the observed events are exemplified in Fig. 9.2. These transition events are mainly caused by two respective cartographic conflicts that drive the generalization of individual buildings. They are densely populated buildings and road networks, which leads to the changes from one pattern type to another. Note that the multiplicity of the transition events is at the pattern level rather than at the building level, indicating one- or many-to-one relationships between the pattern instances. The transition events and their possible causal factors are described as follows.

Event 1 represents the most common transitions (1:1 relationship) where an initial pattern matches to a generalized pattern of the same type. For example, an align-along-road pattern can be matched to another one of the same type (1B in Fig. 9.2). This can be the same for curvilinear patterns (2A in Fig. 9.2) and unstructured clusters (3B in Fig. 9.2).

Event 2 can be observed in situations where an unstructured cluster became a grid pattern after the generalization (e.g. 3A in Fig. 9.2). Events 3–5 describe the areas where the density of buildings enclosed by roads is relatively too high. Since major roads were kept during the generalization, some of the building patterns are

Table 9.1 Observed transition events of building pattern preservation at scale transitions

Event	1:10 k 1:50 k	Multiplicity	1:50 k 1:100 k	Examples shown in Fig. 9.2
1.	Each type of patterns	1:1→	The same type	1B; 2A; 3B
2.	Unstructured cluster	1:1→	Grid pattern	3A
3.	Unstructured cluster	1:1→	Linear alignment	4B
4.	Grid pattern	1:1→	Collinear alignment	N/A
5.	Linear alignments	n:1→	Linear alignment	4A
6.	Linear alignments	n:1→	Grid pattern	4B
7.	Linear alignments	n:1→	Unstructured cluster	N/A
8.	Nonlinear clusters	n:1→	Nonlinear cluster	4C
9.	Align-along-road pattern	1:1→	Collinear/curvilinear pattern	2A

typified by a row of collinear alignment due to limited space and enlarged symbols. To be specific, Events 2–4 show 1:1 relationships between initial and generalized patterns (i.e. a whole pattern is converted into another pattern); while Event 5 (4A in Fig. 9.2) illustrates a n:1 relationship with n $\geq$ 2, where several patterns are typified by one representative pattern); Event 4 describes that a grid pattern can become a collinear alignment but not a curvilinear pattern after the generalization.

Events 6–8 demonstrate the cases where the change of patterns is mainly caused by the removal of road features. This usually occurs in urban areas where the road network is densely populated and minor roads are eliminated during the generalization. In such cases neighboring building patterns may be aggregated or typified into bigger ones and n:1 relationships are usually observed (e.g. 4B in Fig. 9.2). Note that Event 8 (e.g. 4C in Fig. 9.2) may sometimes be difficult to distinguish from Event 6 and 7 as it is sometimes difficult to strictly distinguish between different pattern types.

Align-along-road patterns are somewhat special, as they look like collinear and curvilinear patterns except for their relationships to roads. Event 9 describes a situation where align-along-road patterns change with the removal of the aligned roads, which leads to the change of the patterns into collinear or curvilinear alignments.

All the above transition events can be observed for scale transitions from10 k to 50 k and 50 k to 100 k in our data. The observed events formalize the multiplicity relationships and the changes from one pattern type to another that still ensure the preservation of important characteristics of building patterns. Although these transition events are empirically observed and are subject to further refinement, transitions from e.g. grid to unstructured patterns are considered to be unacceptable. As a result, they can be used to further enrich the written specifications for automatically generalizing and evaluating building patterns.

Practically, it is meaningless to formulate strictly the distinction between the events, mainly because uncertainties exist in distinguishing and detecting different types of patterns. A building pattern can be regarded at the same time as an unstructured cluster, a grid pattern, or multiple rows of linear alignments; an align-along-road type can also be a linear alignment. However, this does not influence the matching process because potential transitions from one type into another are formulated and impossible ones are excluded (Table 9.1). For example, 2A in Fig. 9.2 can be viewed as either 'curvilinear to curvilinear' (Event 1) or 'align-along-road to curvilinear' transition (Event 9); 4B can be described either as three unstructured clusters changing into three rows of linear alignments (Event 3) or as six rows of linear alignments changing into one grid pattern (Event 6).

9.3.3 Method for Automatic Matching Process

Different from previous studies on matching road networks (Devogele et al. 1996; Mustière and Devogele 2007; Zhang and Meng 2007) which make extensive use of topological information to match complex constructs like interchanges and roundabouts, the matching of building patterns mainly focuses on geometric and structural aspects. Besides, since the purpose of this work is to enable automated evaluation of building patterns, the matching of feature attributes (schema matching) is outside the scope of this research.

The matching process consists of two sub-processes: geometric and characteristic matching. The matching is based on two kinds of previously enriched information, namely the pattern descriptions and the transition events. The pattern descriptions include pattern type information, representational geometries and homogeneity properties (Sect. 9.2). The geometries of the pattern instances are used in the geometric matching process. The transition events are used in the characteristic matching process. When a generalized pattern is geometrically matched with an initial pattern, type information of the two patterns is checked with respect to the list of transition events; if the type information is consistent with the transition events, the matching pair is selected as a candidate for further matching; if otherwise the matching of the two fails. In this latter case, one can conclude that the pattern is not preserved during generalization. It is worth noting that the buildings to be evaluated are partitioned by road networks and each step of the automatic matching is restricted to buildings within a partition cell. The technical detail of the process is described as follows.

The geometric matching deals with the similarity between geometries of the patterns to be matched. In the matching of the same type patterns (event 1), the similarity can be measured by distances like nearest distance, Hausdorff distance, and Fréchet distance (Alt and Godau 1995; Mascret et al. 2006). To match polygons (nonlinear patterns), another similarity measure (i.e. contrast model) developed by Tversky (1977) is appropriate. In the cases events 3–7 where linear alignments are matched with nonlinear clusters or several linear alignments are matched to one linear alignment, buffers of skeletons should be used instead of skeletons alone for representing linear alignments to improve the matching result. In our first experiment (Section 9.4.1), we used nearest distance and the contrast model as similarity measures to match between linear and polygonal representational geometries respectively, for simplicity reasons.

The result from geometric matching can be refined by characteristic matching, especially when the result contains several candidates. When matching the patterns of the same type, the characteristics like pattern type, align angle, curvature and aligned road are of great importance. For example, Fig. 9.3a describes a possible case where a generalized collinear alignment (the gray buildings) is (potentially) matched with two collinear patterns (A1 and A2) and a curvilinear one, as they are equally close to each other (d1 = d2 = d3 in Fig. 9.3b). Such an ambiguous situation can be caused by displacing A1 to avoid the widened road symbol. To simplify the situation, pattern type is used to filter out the curvilinear pattern A3, and then the angle of the skeletons (S1, S2, S3, and GS) is calculated. Since the angle between GS and S1 is the smallest, the process matches the generalized alignment to A1 and the matching process is terminated for this generalized alignment.

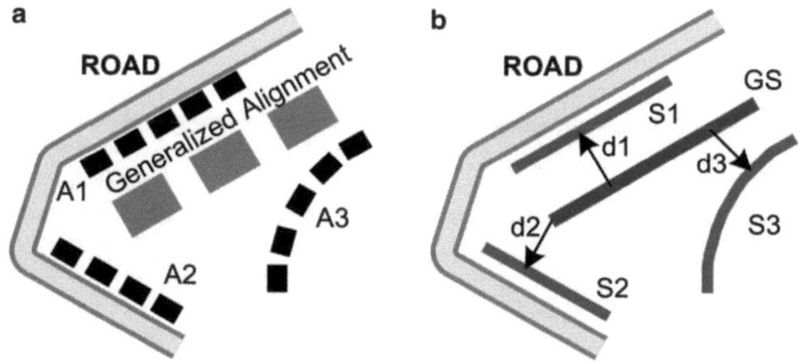

Fig. 9.3 Automated matching of building patterns: (**a**) matching situation and (**b**) geometric and characteristic matching process

9.3.4 Automated Evaluation and Interpretation of the Evaluation Results

9.3.4.1 Evaluation by Computing Changes Between Matched Building Pattern Objects

After all the required information has been collected (i.e. detected building patterns, transition events, and automatically matched patterns), the automated evaluation is carried out by firstly measuring the characteristics defined in Sect. 9.2 for each building pattern in a matching pair, and then comparing the measured characteristics between the patterns. The interpretation of the comparison is described in Sect. 9.3.4.2.

To be more concrete, the characteristics measured for both initial and generalized building patterns could be a set of values (e.g. homogeneities of spacing, sizes, etc.) or an aggregated value (summary the previous values). In this paper we implement the evaluation based on separate values as a proof-of-concept. Note also that if matched patterns are of the same type, then all their characteristics are also matched, and then the evaluation is simply comparing their measured homogeneity values in a pair-wise manner; if they are of different types, only 'basicHomo' is compared. In the cases where n ($n \geq 2$) initial patterns are linked to one generalized pattern, the initial homogeneities should be summarized by the weighted average method; the weighting is based on lengths of skeletons (linear alignments), and on areas of convex hulls (nonlinear patterns).

9.3.4.2 Interpret the Quantitative Evaluation Results

The next key issue of the automated evaluation of building pattern preservation is to decide the acceptable values for the pattern characteristics after the generalization, based on which the quantitative evaluation results can be qualified into a human readable format (e.g. 'acceptable' or 'unacceptable'). According to Bard and Ruas

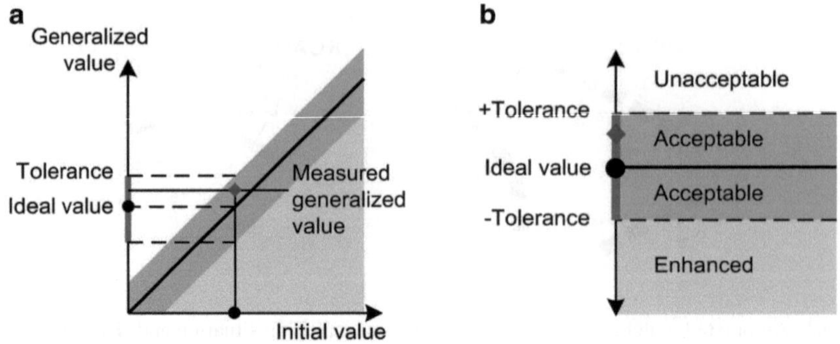

Fig. 9.4 Idealized evolution function (**a**) and interpretation function (**b**)

(2004) an idealized evolution function can be specified for each preservation constraint, where target characteristic values should be equal to initial values (Fig. 9.4a). In allow for the flexibility as in manual generalization, a tolerance is introduced (dark gray areas in Fig 9.4).

In our research, we slightly modify the interpretation function proposed by Bard and Ruas (2004) to be more appropriate for the pattern preservation constraint. That is, if a measured homogeneity falls into the dark gray area in Fig. 9.4b (|MeasuredGenVal – IdealVal| $\leq$ Tolerance), then this property is considered as being preserved (marked as 'acceptable'). This is motivated by the fact that small deviations are tolerable to human eyes. If the value is larger than 'ideal value' by a unit of tolerance, the property is regarded as not being preserved ('unacceptable'); while if the value is less than 'target value' by a unit of tolerance (light gray area in Fig. 9.4b), the property is regarded to be 'enhanced'. Since when homogeneity decreases a pattern becomes more regular (enhanced). This enhancement is preferred in map generalization.

9.4 Implementation and Validation of the Methodology

This section implements and validates the methodology presented in Sect. 9.3 by applying it to a test case. The test case consists of two datasets different from the datasets used for the visual analysis (Sect. 9.3): one is a Dutch topographic dataset at 1:10 k (Fig. 9.5a); another dataset (TOP50NL) is interactively generalized from the TOP10NL (Fig. 9.5b). The interactively generalized dataset is assumed as a good generalization example. The hypothesis of this experiment is that, by applying the methodology to the test datasets the evaluation result should confirm the fact that the building patterns in the data are preserved sufficiently. This is what can be expected from the interactively generalized data. In the following, we present the results of pattern matching (Sect. 9.4.1) and validation (Sect. 9.4.2).

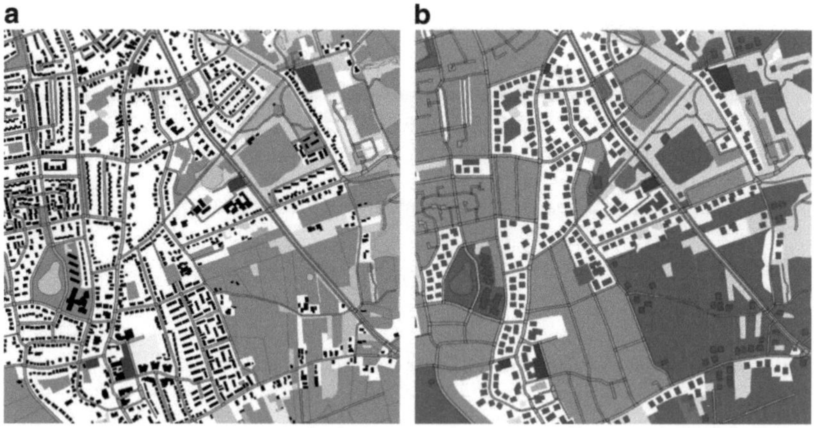

Fig. 9.5 Test datasets for validating our evaluation methods: (**a**) TOP10NL and (**b**) TOP50NL

9.4.1 Pattern Matching Result

A previous work on pattern detection and characterization was applied to the test datasets. These detection results have been reported in Zhang et al. (2010). After that, we applied the automated matching process to match the detected patterns. The matching result is as follows. In most cases, one partition cell contains only one to two detected patterns, so around 94% of the patterns (48 out of 51) detected in TOP50NL were correctly matched to their correspondences in TOP10NL. There are 53% of the patterns (68 out of 129) detected in TOP10NL that are mismatched. Most of the mismatch is because the composing buildings have been transformed into built-up areas (light red areas in Fig. 9.5b). The remaining mismatch is due to the absence of corresponding patterns in TOP50NL, in which building patterns may have been removed to satisfy other important constraints (e.g. minimum distances). There are also cases where some patterns (e.g. unstructured clusters) were correctly matched with others of different types (e.g. collinear alignments) and where n:1 relationships were correctly created.

9.4.2 Validation

The homogeneities were calculated for all matched patterns, and some rather simple examples (1:1 relationship) are illustrated (see e.g. Fig. 9.6 and Fig. 9.7). Instead of using standard deviation (STD), we calculated the homogeneities using coefficient of variance (STD/Mean), since coefficient of variance is superior to standard deviation in characterizing the homogeneity of building patterns (spacing, size, and shape). Homogeneity of orientation, however, should be computed by

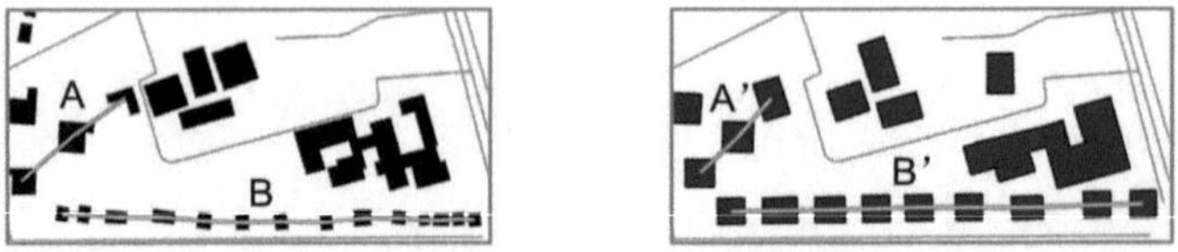

ID	Spacing (std/mean)	Size (std/mean)	Orientation (std)
A	0.06	0.32	9.74°
A'	0.02	0.11	8.99°
B	0.52	0.26	5.57°
B'	0.35	0.09	0.00°

Fig. 9.6 Changes of the characteristics of linear patterns at a scale transition

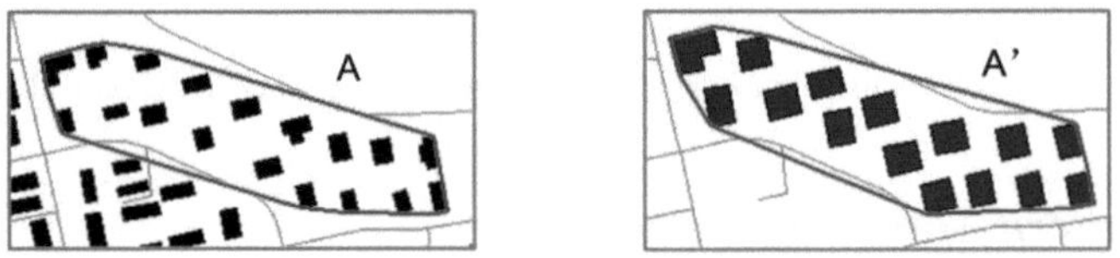

ID	Spacing (std/mean)	Size (std/mean)	Orientation (std)
A	0.22	0.22	2.03°
A'	0.19	0.19	1.17°

Fig. 9.7 Changes of the characteristics of unstructured clusters at a scale transition

STD, as it is meaningless to calculate coefficient of variance for a cyclic variable (orientation).

The result shows that in general the detected patterns were enhanced during the generalization because the homogeneities of spacing, sizes, and orientations were reduced. This shows the proposed methodology for automated evaluation of building pattern preservation has the potential in practice. In a second step regression functions indicating the relationships between initial and target values were established, where the values reflect the homogeneities of spacing, sizes, and orientations (Fig. 9.8). But currently, these regressed functions should not be used as final references, as the sample size used in this experiment was limited.

The result in Fig. 9.8 confirms that the homogeneity values of the patterns are significantly reduced (most of the data points are below the ideal preservation function in Fig. 9.8). This means that the patterns become more regular. To be more specific, most of the preserved building patterns become more regular in terms of spacing, size, orientation, and shape. The main reason of the increase in regularity is that the building pattern preservation constraint keeps the structural aspects (i.e. regularity) of the patterns. On the other hand, legibility constraints also affect the sizes and shapes of all buildings. That is, minimum size constraint has a threshold effect that will lever the difference of sizes, and minimum dimension constraints simplify building geometries (i.e. reduce shape complexity).

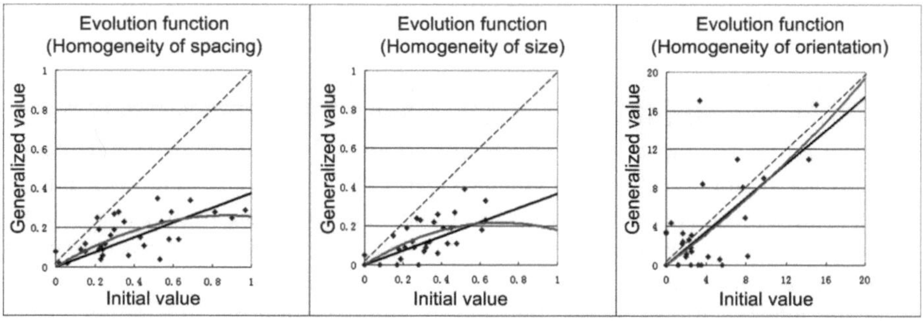

Fig. 9.8 Evolution functions obtained from the test datasets; *black line* – linear function; *red curve* – nonlinear function; *dotted line* – ideal preservation function

However, the homogeneities of orientation of the building patterns were not significantly reduced. About half of the data points in Fig. 9.8 (left) are above the idealized function, and most of the deviations are less than 5°. Considering the fact that angular deviations from 0° to 10° are less discernable by human eyes, the possibility of introducing the deviations in the interactive generalization is high. Therefore, the result in Fig. 9.8 left is regarded as 'the regularity of orientation being preserved'. To incorporate such flexibility for a better evaluation, a tolerance should be added. The deviations may have other reasons. In some cases, some buildings, which are not part of a pattern in TOP10NL, become part of it in TOP50NL; these buildings contribute a lot to the increase in the homogeneity of orientation. In other cases, some buildings in a pattern were displaced and rotated to meet the minimum distance constraint, which also increases the homogeneity of orientation.

9.5 Discussion and Conclusions

This paper studies the supported techniques for automated evaluation of the building pattern preservation constraint. The automated evaluation of this constraint is currently impeded by issues concerning pattern detection, pattern matching, and constraint formalization. We address these issues by firstly summarizing previously obtained results in pattern detection and characterization, on which the pattern matching and automated evaluation are based. The classification and characterization of building patterns formalizes our knowledge on building patterns. The proposed methodology also contributes to the automated evaluation of preservation constraints in general.

We find further that the matching of building patterns is not straightforward because one type of pattern may become another type. As a result, acceptable changes (transition events) for building patterns are identified and formulated by studying existing maps. We suggest that these events should be used to enrich the written specifications for automated processes. Subsequently, a two-step matching process is proposed based on the enriched pattern descriptions and the transition events.

The presented methodology are implemented and validated by being applied to interactively generalized data. The major result shows that the methodology has the potential in the designated task. The results confirm also that the pattern detection and characterization methods in Zhang et al. (2010) are consistent with human perception; the detection method proves to be also applicable to non-Dutch topographic maps as the method was successful in detecting building patterns in maps of France, Spain, and China.

In the interpretation process, the preservation function and tolerances in our current approach can be predefined by experts or specified by users. The evolution functions used in this paper appear to be an oversimplified form. Burghardt and Schmid (Schmid 2008; Mustière et al. 2009) have shown a reverse engineering technique to fit the evolution functions, which is similar to the validation approach (regression analysis) used in this paper. Although this approach may be more adaptive to the training data, a more complex function depends largely on the training sample and may also cause problems such as over-fitting, which is not preferred in a more generic approach.

Currently, the proposed evaluation approach only focuses on the building patterns that have been kept, and in this respect the approach was effective and sufficient. However, there still lacks of a more generic framework in which more constraints and measures should be introduced to be also able to evaluate the building patterns that have been diminished or transformed to built-up areas. Additionally, the implementation used separate characteristics to assess the change of building patterns. This is not straightforward for the system to make evaluation decisions. For a more practical approach, future work should evaluate the patterns by aggregating the separate characteristics into a single value to make meaningful decisions. Finally, future work can integrate the proposed evaluation approach with the evaluation of other constraints to obtain overall qualities, indicating acceptable generalization solutions.

Acknowledgments This work was partly supported by the 863 Program (Grant No. 2009AA121404). Faculty of Geo-Information Science and Earth Observation (ITC) of the University of Twente, which funds the first author as a PhD student, is gratefully acknowledged.

References

Alt H, Godau M (1995) Computing the Fréchet distance between two polygonal curves. Int J Comput Geom Appl 5(1–2):75–91

Anders KH (2006) Grid typification. In: Riedl A, Kainz W, Elmes GA (eds) Progress in spatial data handling. Springer, Vienna, pp 633–642

Bard S, Ruas A (2004) Why and how evaluating generalised data? In: Developments in spatial data handling (SDH'04), Spring, Leicester, pp 327–342

Bobzien M, Burghardt D, Petzold I, Neun M, Weibel R (2008) Multi-representation databases with explicitly modeled horizontal, vertical, and update relations. Cartogr Geogr Inform Sci 35(1): 3–16

Burghardt D, Schmidt S, Stoter J (2007) Investigations on cartographic constraint formalisation. In: 10th ICA workshop on generalisation and multiple representation, Moscow. Available at http://aci.ign.fr/BDpubli/moscow2007/Burghardt-ICAWorkshop.pdf

Devogele T, Trevisan J, Raynal L (1996) Building a multi-scale database with scale-transaction relationships. In: Kraak MJ, Molenaar M (eds) Proceedings of the 7th international symposium on spatial data handling (SDH'96), Taylor & Francis, Delft, pp 337–351

Hampe M, Anders K, Sester M (2003) MRDB applications for data revision and real-time generalization. In: Proceedings of the 21st international cartographic conference, Durban, pp 192–202

Lüscher P, Weibel R, Burghardt D (2009) Integrating ontological modelling and Bayesian inference for pattern classification in topographic vector data. Comput Environ Urban 33: 363–374

Mackaness WA, Ruas A (2007) Evaluation in the map generalisation process. In: Mackaness WA, Ruas A, Sarjakoski LT (eds) Generalisation of geographic information: cartographic modelling and applications. Elsevier, Amsterdam, pp 89–111

Mascret A, Devogele T, Berre IL, Hénaff A (2006) Coastline matching process based on the discrete Fréchet distance. In: Riedl A, Kainz W, Elmes GA (eds) Proceedings of the 12th international symposium on spatial data handling, Vienna, pp 383–400

Mustière S, Devogele T (2007) Matching networks with different levels of detail. Geoinformatica 12(4):435–453

Mustière S, Sester M, van Harmelen F, van Oosterom P (2009) 09161 Abstracts collection – generalization of spatial information. In: Mustière S, Sester M, van Harmelen F, van Oosterom P (eds) Generalization of spatial information. Schloss Dagstuhl – Leibniz-Zentrum fuer Informatik, Germany

Regnauld N (1996) Recognition of building clusters for generalization. In: Kraak MJ, Molenaar M (eds) Advances in GIS research II (Proceedings of 6th SDH'96, Delft), Taylor & Francis, London, pp 4B.1–4B.14

Ruas A, Holzapfel F (2003) Automatic characterisation of building alignments by means of expert knowledge. In: Proceedings of the 21st international cartographic conference (ICC), Durban, South Africa, 10–16 Aug 2003, pp 1604–1615

Schmid S (2008) Automated constraint-based evaluation of cartographic generalization solutions. Unpublished MSc thesis, Department of Geography, University of Zurich

Stoter J et al (2009a) Methodology for evaluating automated map generalization in commercial software. Comput Environ Urban 33:311–324

Stoter J, van Smaalen J, Bakker N, Hardy P (2009b) Specifying map requirements for automated generalization of topographic data. Cartogr J 46(3):214–227

Tversky A (1977) Features of similarity. Psychol Rev 84:327–352

Weibel R, Dutton GH (1998) Constraint-based automated map generalization. In: Poiker, TK, Chrisman N (eds) Proceedings of the 8th international symposium on spatial data handling, Vancouver, pp 214–224

Zhang M, Meng L (2007) An iterative road-matching approach for the integration of postal data. Comput Environ Urban 31(5):597–615

Zhang X, Ai T, Stoter J (2010) Characterization and detection of building patterns in cartographic data. In: Joint international conference on theory, data handling and modelling in GeoSpatial information science, Hong Kong, vol XXXVIII, Part 2, pp 261–266

Part IV
Pattern Recognition and Object Extraction

Chapter 10
Graph Based Recognition of Grid Pattern in Street Networks

Jing Tian, Tinghua Ai, and Xiaobin Jia

Abstract Pattern recognition is an important step in map generalization. Pattern recognition in a street network is significant for street network generalization. A grid is characterized by a set of mostly parallel lines, which are crossed by a second set of parallel lines roughly at right angles. Inspired by object recognition in image processing, this paper presents an approach to grid recognition in street network based on graph theory. Firstly, bridges and isolated points of the network are identified and repeatedly deleted. Secondly, a 'similar orientations' graph is created, in which the vertices represent street segments and the edges represent similar orientation relationships between streets. Thirdly, the candidates are extracted through graph operators such as the finding of connected components, finding maximal complete sub-graphs, joins and intersections. Finally, the candidates are repeatedly evaluated by deleting bridges and isolated lines, reorganizing them into stroke models, changing these stroke models into street intersection graphs in which vertices represent strokes and edges represent strokes intersecting each other. The average clustering coefficient of these graphs is then calculated. Experimental results show that the proposed approach is valid in detecting the grid pattern in lower degradation situations.

J. Tian (✉) • T.H. Ai • X.B. Jia
School of Resource and Environment Science, Wuhan University, 129 Luoyu Road, Wuhan 430079, China

Key Laboratory of Geographic Information System, Ministry of Education, Wuhan University, 129 Luoyu Road, Wuhan 430079, China
e-mail: yutaka-2010@163.com

A. Yeh et al. (eds.), *Advances in Spatial Data Handling and GIS*, Lecture Notes in Geoinformation and Cartography,
DOI 10.1007/978-3-642-25926-5_10,

10.1 Introduction

> Ideas of scale and pattern are central to the process of interpretation in the geosciences. The patterns that are evident at any given scale will have specific causes and consequences (Mackaness 2007, pp. 1–2).

What is pattern and pattern recognition? Traditionally, pattern recognition is the scientific discipline, the goal of which is the classification of objects into a number of categories or classes. Depending on the application, these objects can be images or signal waveforms or any types of measurements that need to be classified (Theodoridis and Koutroumbas 2009). The "pattern" refers to objects. From the geosciences experts' view, patterns can be organized into partonomic hierarchies reflecting relationships between parts and subparts (Mackaness and Edwards 2002). In summary, a pattern can be an object, object cluster, property with an object or object cluster, and relationships between objects.

Pattern recognition techniques have been applied to machine vision, character recognition, and computer-aided diagnosis. These applications use images as information sources and benefit from recognition of specific patterns in images. In a raster mode, the main tasks of pattern recognition are object detection and features extraction. Here, streets must be extracted from an image and this work is the focus. In vector mode, however, the elementary objects points, lines and polygons already exist as a spatial database. Streets already exist in geographical spatial databases and the recognition of relationships and structures between and among streets is the interest in this study. Traditional pattern recognition is little concerned with this type of problem.

In recent years, pattern recognition of object spatial clusters has come to assume increasing importance in the map generalization field. The relationship between pattern recognition and map generalization is that pattern recognition is the starting point for map generalization, and pattern maintenance is the aim of map generalization. Maintaining patterns after generalization is not only a generalization guideline but also the result of a good generalization process..

Identifying and extracting street network spatial patterns is significant for network generalization. Maintaining the pattern of the street network after generalization is the basic principle in the selection of streets. The pattern here refers to the structure of the street network. However, most street network generalization methods pay no attention to the principle emphasized by Zhang (2004) as one of "modeling the patterns as input to generalization algorithm".

Perceptual organization refers to the ability of human beings to form meaningful groupings of low level tokens detected in an image (Sarkar and Boyer 1994). Sarkar and Boyer (1994) present a hierarchical approach to perceptual organization for detecting structure among tokens. Tokens are perceptual objects which are to be organized. The tokens may be points, straight lines, curves, parallel lines and rectangles, etc. The tokens at each level have specific attributes, and some relationship graphs where nodes represent tokens and edges represent the associations between tokens are created using these attributes. For example, a proximity graph has links joining nodes which represent tokens having points close together. Tokens at high level are extracted from tokens at low level by standardized graphing operations.

Inspired by Sarkar and Boyer's idea (1994), this paper presents an approach to recognition of grid patterns in street networks. A grid is characterized by a set of roughly parallel lines, which are crossed by a second set of parallel lines at a roughly perpendicular angle. A set of parallel lines consists of line segments having proximity and a similar orientation. The parallel lines and the grid are extracted by graph theory operators such as finding connected components, finding maximally complete sub-graphs, joins and intersections. The rest of this paper is organized as follows. Section 10.2 is a review of work on the recognition of patterns in street networks. Section 10.3 introduces concepts and definitions in graph theory which are used in this paper. Section 10.4 presents the proposed approach to the recognition of grid patterns in detail. An experiment was conducted in order to test the approach and its advantages and disadvantages are discussed in Sect. 10.5. Finally, the concluding Sect. 10.6 offers several avenues for future work.

10.2 Related Work

10.2.1 Main Patterns in Street Networks

Street networks consist of a large number of streets which intersect in different modes. The patterns can be classified according to different criteria. From a geometric view, in Zhang's research (2004), discernable patterns in road networks are star-like, grid-like and irregular. Marshall (2005) proposed that the main patterns have linear, treelike, radial, cellular and hybrid forms. Heinzle et al. (2006) suggested the typical patterns are strokes as a type of linear form, grids as a type of cellular form, stars as a type of radial form and circular roads as a type of cellular form. It is important to note that the grid pattern in Heinzle et al. (2005) only consists of streets which are parts of a grid. The grid is defined in a local sense. From the topological view, there exists the three important patterns of regular, small-world and random (Watts and Strogatz 1998).If the criterion is the distribution of the node degrees of orientation in the network, a random and scale-free pattern can be used in such classification. Xie and Levinson (2007) defined four typical connection patterns in road networks. They are Ring and Web forms in circuit networks and Star and Hub-and Spoke forms in branching networks. A circuit is defined as a closed path that begins and ends at the same vertex. Branching networks consist of sets of connected lines lacking any complete circuit.

10.2.2 Characteristics and Quantification Measures of Patterns

A grid is characterized by a set of roughly parallel streets, which are crossed by a second set of parallel streets at a roughly perpendicular angle. The main quantification measure is the statistical distribution of street directions. A star is characterized by a "fuzzy centre" degree of connectivity from which streets radiate or converge. If the fuzzy centre is clustered to one intersection, its connectivity degree is high compared with other intersections. A ring is approximately a circle. There are rings of different sizes: roundabouts, rings around hub-like junctions, rings surrounding a city centre. The ring can be described by the shape indices compactness, convexity, and geometric moment. To check whether a network is small-world, a characteristic path length and a clustering coefficient are proposed. For a scale-free network, degrees of connectivity and degree distributions are used. The connectivity degree of a node is the number of edges coincident at that node. The degree distribution is defined as the probability distribution of nodes with specific numbers of links. Porta et al. (2006) found that the clustering coefficient of grid patterns is extremely small, indicating that the geometric pattern and topological pattern of street network are intrinsically related.

10.2.3 Methods of Recognition of Patterns

For a grid pattern, Heinzle et al. (2005) select so called CRS nodes, at which four edges intersect, as starting points, investigated the histogram of the edge directions, and then examined the polygon with respect to its similarity with neighboring polygons. Yang et al. (2010) proposed a method for identifying grid patterns. Their method first builds a node-edge topology, generates polygons from these edges, calculates a set of polygon parameters, and finally identifies a grid using a multi-criteria decision method. These two methods for grid recognition both adopt a polygon-based structural recognition approach.

For the star pattern, Heinzle et al. (2005, 2007) take any node in the graph as a potential centre point, use the Dijkstra algorithm to compute the single source shortest path from this node to all other nodes, and then use a circle with a certain radius to search for the centre. A fuzzy centre is determined by a cluster algorithm. The problem of this approach, as they pointed out, is lack of knowledge of the search radius and the range of possible configurations.

For the ring pattern, Heinzle et al. (2006) use convex hull peeling techniques to determine the approximate position of the city centre, neighboring polygons are aggregated as long as they are not subsequently separated by a major stroke, and then some candidates are sought, using geometric moments, curvature and convexity to evaluate the candidates. The problem of this approach is the huge degree of uncertainty in evaluating whether the candidate is a ring.

It is easy to check for a small-world network. The small-world has a small characteristic path length than random graphs and a much larger coefficient than a random graph. For scale-free networks, it is easy to check whether the connectivity degree distribution follows a power law.

10.2.4 Summary

It can be concluded from the above that (1) there has been much work on classification and qualitative description of geometric patterns in street networks, but little work on how to quantify the characteristics and how to detect geometric patterns. (2) Geometric patterns cognitively are easy to recognize but difficult to formalize whereas topological patterns it are difficult to recognize but easy to formalize. The association of various measurements has not been studied.

This study proposes a method for detecting grid patterns. The major difference between the proposed method and the methods of Heinzle et al. (2005) and Yang et al. (2010) is that the new method detects grid patterns through the grouping of street segments.

10.3 Concepts and Definition in Graph Theory

A street network is usually modeled using a graph, where vertices represent intersections and edges represent street segments. This section introduces some basic graph concepts and definitions which are used in the next two sections. For a more complete introduction to graph theory, readers can refer for example to Diestel (2006).

A *graph* of G consists of a finite set of *vertices* (or nodes, or points) and a finite set of *edges* (or links, or lines). A graph is often denoted as $G(V,E)$ where $V = \{v1, v2 \cdots vn\}$ and $E = \{vi, vj\}$. Two vertices vi, vj of G are *adjacent*, if $vivj$ is an edge of G. The number of vertices of a graph G is its *order*. G is an *undirected graph* if E is an unordered set. A *complete graph* $G = (V,E)$ is one in which each vertex is connected to every other vertex. Let $G = (V,E)$ and $G' = (V',E')$ be two graphs. If $V' \subseteq V$ and $E' \subseteq E$, then $G'$is a *subgraph* of G. A *maximal complete subgraph* is a complete subgraph of G that contains as many vertices of G as possible. A non-empty graph is called *connected* if any two of its vertices are linked by a path in G. A maximal connected subgraph of G is called a *connected component* of G.The degree of a vertex is the number of edges at this vertex. A vertex of degree 0 is *isolated.* A vertex of a connected graph is an *articulation vertex* if its removal would create a separate graph. An edge of a connected graph is a *bridge* if deleting it would create a separate graph. We set $G \cap G' := (V \cap V', E \cap E')$, where $\cap$ is *intersection* operation. *Join* is an operation between two graphs that links every vertex of one graph to every vertex of the other graph.

A *relation graph* is a graph in which the vertices represent entities and the edges represent specific relationships among entities. The structural representation of a street network proposed by Jiang and Claramunt (2004) is a relation graph, where vertices represented named streets and edges represent intersection relation among streets. The *street segment intersection graph* is a graph whose vertices represent street segments and edges represent intersection relationships among street segments. The *similar orientation graph* is a graph whose vertices represent street segments and edges represent similar orientation street segments. The orientation of the street segment is defined by its azimuth angle, of range (0°, 180°). The range of angular separation between two street segments is (0°, 90°). The *street intersection graph* is a graph whose vertices correspond to entire streets and edges correspond to intersection relationships among entire streets. The *entire street* consists of street segments which satisfy the "good continuation" rule (Thomson and Richardson 1999).

10.4 Recognition of Grid Pattern in Street Networks

10.4.1 Characterization of Grid

If the element considered is a line, a grid is characterized by a set of roughly parallel streets, which are crossed by a second set of parallel streets at a roughly perpendicular angle. The perpendicular angle is not an essential characteristic of a grid. If the element considered is a polygon, a grid is composed of a set of polygons of similar shape and connection mode. Similar to the viewpoint of that of Heinzle et al. (2005) and Xie and Levinson (2007), in this study a grid is defined as a closed structure consisting mostly of sets of parallel lines, crossed by a second set of parallel lines, not including other lines. Hence, for the recognition of a grid, the key questions are: (1) How can the two sets of parallel lines be found? (2) How can it be evaluated whether the result is a grid or not?

In the perceptual organization, Sarkar and Boyer (1994) chose the length of two lines, perpendicular distance between the mid point of the smaller line and the larger line and the angular separation to describe parallelism. The threshold value for these parameters is decided using statistical methods and experiments. By formulating the perceptual characteristics leading to human recognition of parallelism in terms of several distinctive forces, Ip and Wong (1997) developed a force-driven model as a new optimization strategy to perform correspondence establishment between points in the matching curves. To detect parallel segments after coupling, they first made sure that the orientations fell within an allowable limit. From the analysis above, it can be concluded that the angular separation between two segments is the most important factor for recognition of parallelism.

Porta et al. (2006) found the average clustering coefficient of grid patters to be extremely small. This indicates that the grid pattern and the average clustering coefficient have an intrinsic relationship. In fact, transforming a perfect grid to a street intersection graph (vertices represent entire streets and edges represent the intersection relationships), the value of the average clustering coefficient for such a graph is 0. The value in the experiment of Porta et al. (2006) is not 0 because there exists some disturbance in the streets pattern shown in Fig. 10.1. The grid pattern referred to by Porta et al. (2006) is in the global sense. The average clustering coefficient can be used to evaluate how perfect a grid is.

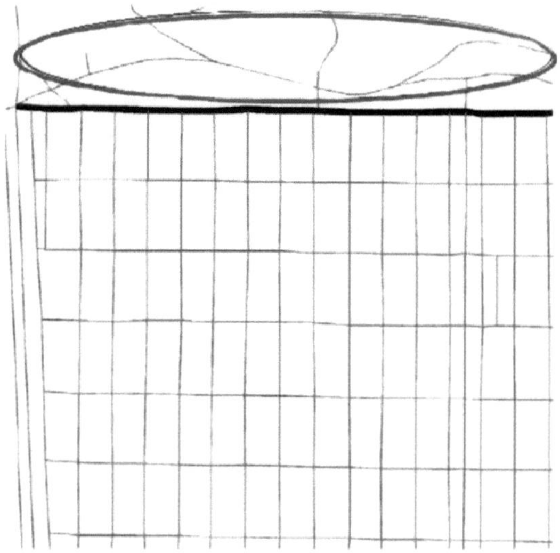

Fig. 10.1 Streets influencing clustering coefficient (modified from Porta et al. 2006)

10.4.2 The Approach to Recognition of Grid Pattern

The core idea is that grids are composed of two groups of parallel streets. There are relationships among the line segments in each group and among the line segments in two groups. The angles of every line segment in the one group are similar, including the parallelism and collinearity. The line segments in one group are intersected by at least one line segment in the other group. Hence, angular similarity is taken as the first compelling condition. The candidates are sough through graph-theoretic operations such as finding connected components, finding a maximal complete subgraph, intersections and joins. Finally, the street intersection graph is formed and the clustering coefficient is computed to evaluate the sets of candidates. The logic of this idea is that all parallel lines would be found and grouped according to their angular separations first and the final result is the subset of these parallel line segments, providing a grid exists. A candidate is filtered through graph-theoretic operations stage by stage. The steps are described in more detail below.

Step 1: Data preprocessing. The street network is organized as a graph in which the vertices represent street intersections and edges represent street segments, i.e. a primal graph. Inspired by Xie and Levinson (2007), the bridges are identified, and deleted and isolated vertices also, until there are no bridges and isolated vertices in the graph. The reason is that a grid contains no bridges and isolated vertices. This step is to clean and refine the street network. Figure 10.2 shows a simple example. Deleting bridge 13 produces Fig. 10.2b, removing the isolated vertex produces Fig. 10.2c,d is a street segment intersection graph of Fig. 10.2c.

Step 2: Street grouping. The graph of similar orientations is formed in which the vertices represent street segments and edges represent street segments having similar orientation. Figure 10.3 shows a similar orientation graph of Fig. 10.2c with two connected components. There exists no grid or else the angular separation

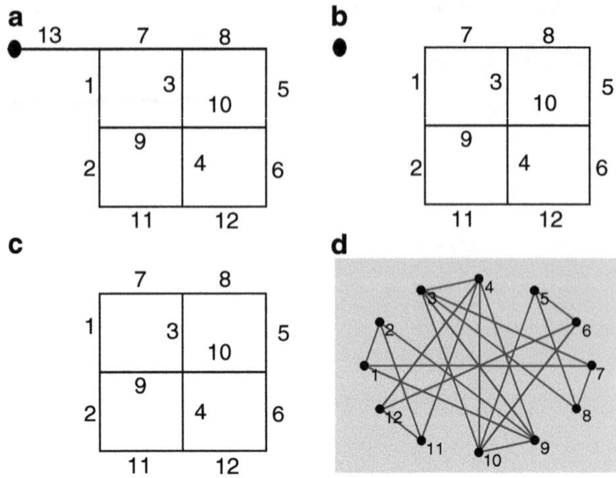

Fig. 10.2 Simple example ((d) visualization by Pajek)

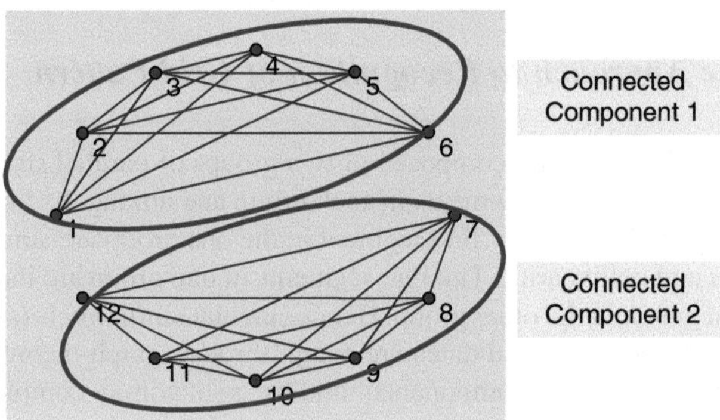

Fig. 10.3 Similar orientation graph with two connected component (visualization by Pajek)

is set too large, if there is only one connected component in a similar orientation graph. For example, if the angular separation is 90°, of course there is only one connected component. The intention of this step is to extract those street segments which are likely to form the grid maybe including many street segments.

Step 3: Forming sets of parallel streets. For each connected component in the similar orientation graph, the maximal complete subgraph is found. The purpose of this step is to group lines of similar orientation in the one set, including parallelism and collinearity. This is based on the truth "line segment angles in one group of parallel lines are similar to each other." This step eliminates the streets which are not parts of a grid. Figure 10.4 shows the maximal complete graph of two connected components in Fig. 10.3. The Maximal complete graph is the same as the connected component because the connected component is itself a complete graph. See Sect. 5.1 for the general case.

Step 4: Linking two sets of parallel streets to form candidates. Select two maximal complete subgraphs, "Join" them, then conduct an "intersection" operation

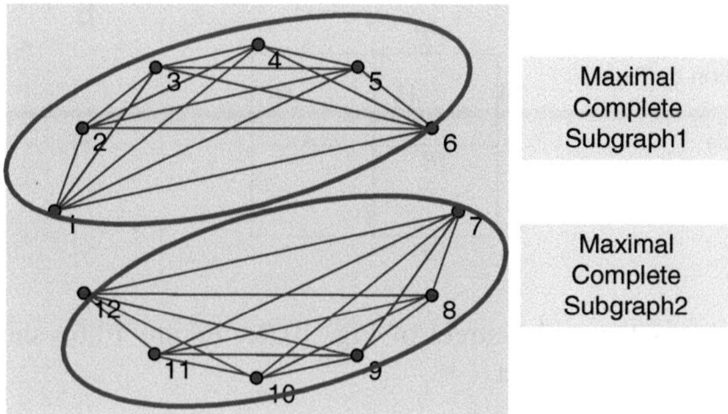

Fig. 10.4 Maximal complete subgraph (visualization by Pajek)

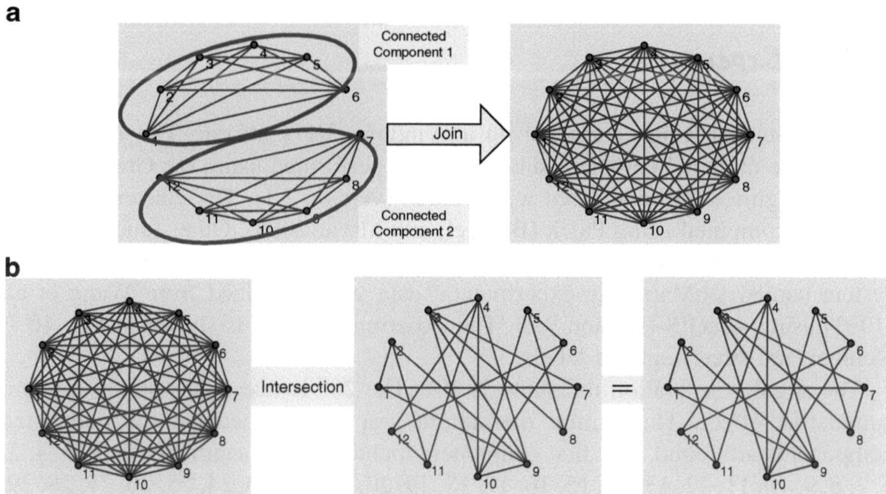

Fig. 10.5 (**a**) "Join" and (**b**) "Intersection" (visualization by Pajek)

with the street segment intersection graph. Figure 10.5 shows the "Join" and "Intersection" operation.

Step 5: Data post-processing. Delete the bridges and isolated vertices until there are no bridges and isolated vertices in the street network after processing by step 4. This step takes into account the closed property of a grid. Figure 10.2c shows a candidate of Fig. 10.2a. See Figs. 10.10 and 10.11 for a complex example.

Step 6: Evaluation. For each candidate, the street segments are linked to form entire streets using the "good continuation" rule (Thomson and Richardson 1999), then the street intersection graph is created. If the order of the graph is less than 6, then the candidate is deleted. The reason is that the minimal grid contains 6 entire streets. The average clustering coefficient of this graph is computed. The candidate is marked out as the final result if its average clustering coefficient value is 0.

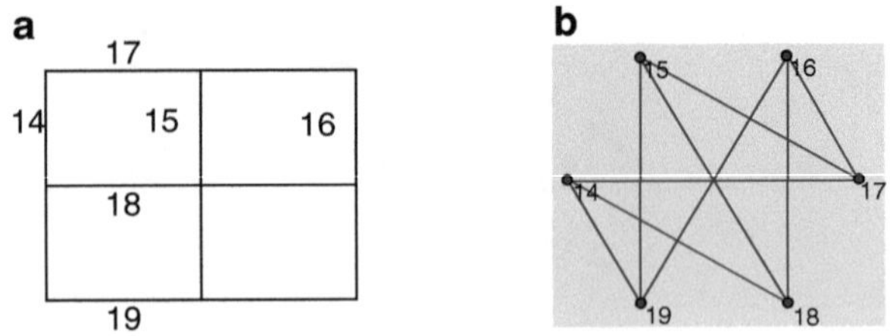

Fig. 10.6 (**a**) Entire streets and (**b**) Street intersection graph (visualization by Pajek)

Figure 10.6a shows the entire street of Fig. 10.2c. Figure 10.6b shows the street intersection graph of Fig. 10.6a.

10.5 Experiments and Discussion

10.5.1 Experiment

The experiments were conducted on a Windows XP platform. The function of creating the maximal complete sub-graph was performed using the Graph Theory Toolbox (Iglin 2003) associated with Matlab (version 7.1). The clustering coefficient was computed using Pajek (Batagelj and Mrvar 2008). Other functions were implemented in Visual C++ (Version 6.0) and integrated into a map generalization system named DoMap. The experimental data were digitized from Wang et al. (1993) using ArcGIS (version 9.2). The experimental data is shown in Fig. 10.7. Numbers are street segment labels.

The angular separation threshold is 20°. The 20° is due to the street segment cumulative effect. The similar orientation graph is formed and its connected components are found. The first component includes street segments: 1, 2, 3, 4, 5, 6, 7, 8, 9, 10, 11, 12, 13, 14, 15, 16, 17, 18, 19, 20, 21, 22, 23, 24, 25, 26, 27, 28, 29, 30, 31, 32, 33, 34, 41, 42, 43, 56, 79. The maximal complete subgraph includes street segments: 1, 2, 3, 4, 5, 6, 7, 8, 9, 10, 11, 12, 13, 14, 15, 17, 18, 19, 20, 21, 22, 23, 24, 27, 28, 29, 30, 31, 32, 33, 34, 56, 79. The corresponding street segment is shown in Fig. 10.8.

The second component includes street segments: 35, 36, 37, 38, 39, 40, 44, 45, 46, 47, 48, 49, 50, 51, 52, 53, 54, 55, 57, 58, 59, 60, 61, 62, 63, 64, 65, 66, 67, 68, 69, 70, 71, 72, 73, 74, 75, 76, 77, 78. The maximal complete subgraph includes street segments: 35, 36, 37, 38, 39, 40, 45, 50, 51, 52, 53, 54, 55, 57, 58, 59, 60, 61, 62, 63, 65, 66, 67, 68, 69, 70, 71, 72, 73, 74, 75, 78. The corresponding street segment is shown in Fig. 10.9.

We next "Join" the two maximal complete subgraph, then "intersect" with the street segment intersection graph created from experimental data in Fig. 10.7. The result is shown in Fig. 10.10.

A candidate is produced after deleting bridges and isolated points until there are no bridges and isolated vertices, as show in Fig. 10.11.

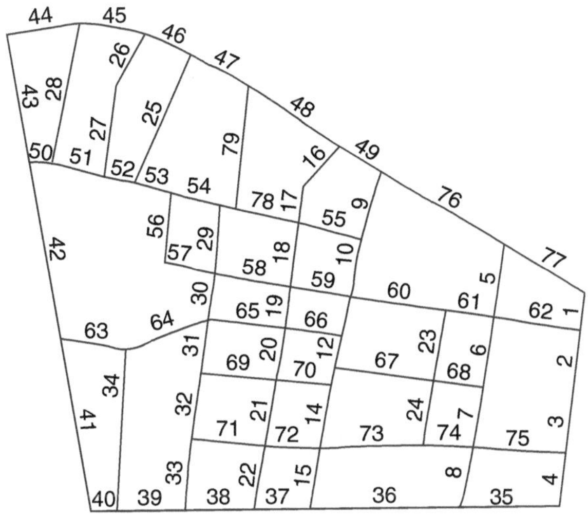

Fig. 10.7 Experimental data

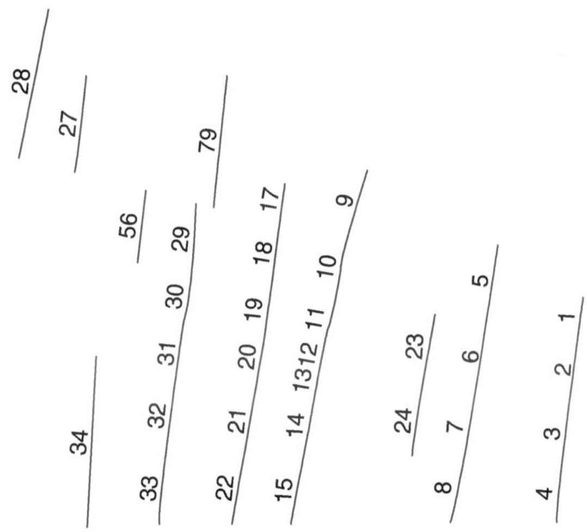

Fig. 10.8 Street segments corresponding to first maximal complete subgraph

The candidate is processed according to the "good continuation" rule to form entire streets, as show in Fig. 10.12a.The street intersection graph is created (Fig. 10.12b), and the average clustering coefficient of this graph is 0.The candidate is marked as the final result.

10.5.2 Discussion

The experimental results show that the proposed approach is valid in detecting the grid pattern in regular and lower degradation situations. The advantages of the proposed approach are: (1) It uses graph-theoretic operations to represent human

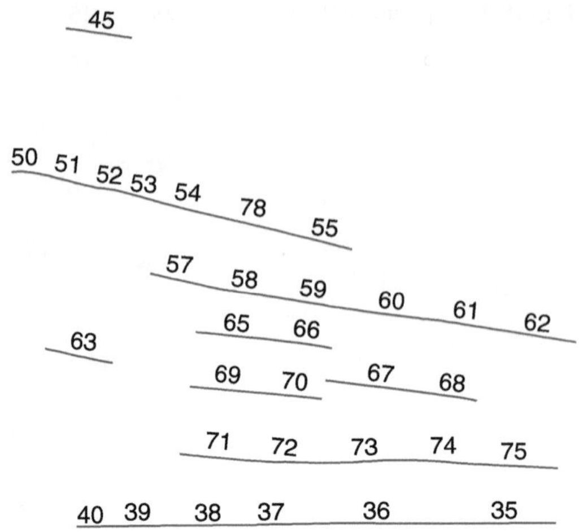

Fig. 10.9 Street segments corresponding to second maximal complete subgraph

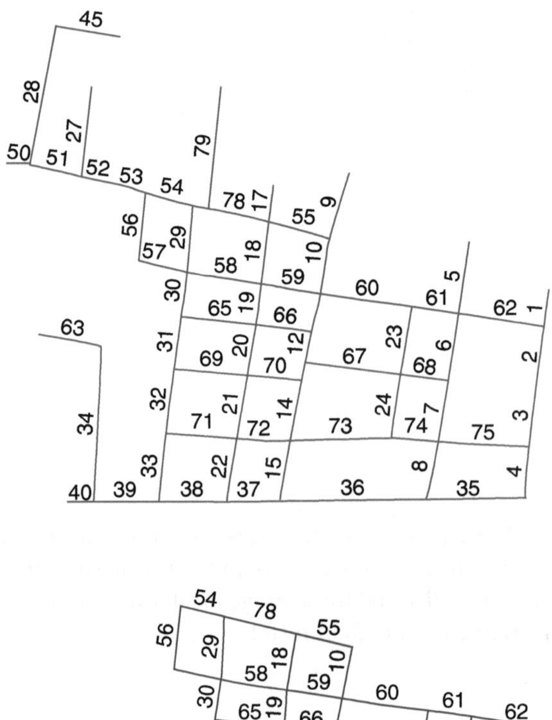

Fig. 10.10 Street segments after join and intersection

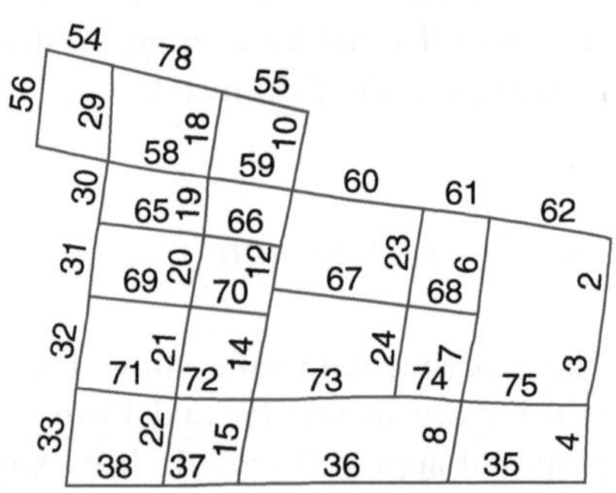

Fig. 10.11 A candidate

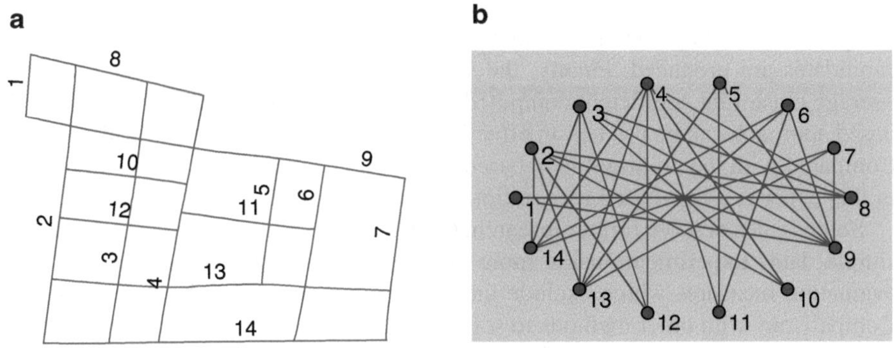

Fig. 10.12 (**a**) Entire streets and (**b**) street intersection relation graph (visualization by Pajek)

Fig. 10.13 One exception case

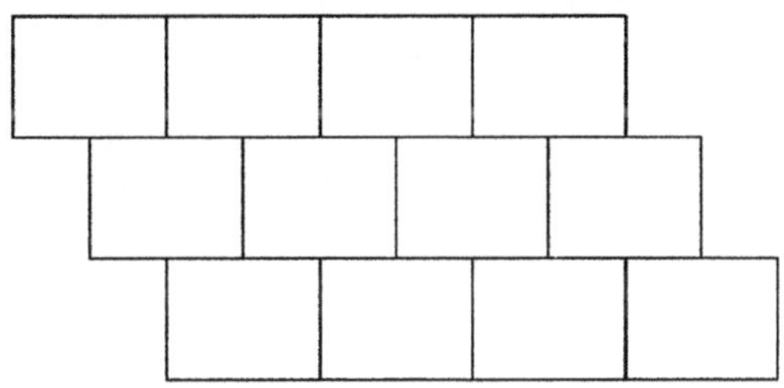

recognition knowledge. i.e. finding a set of roughly parallel streets, which are crossed by a second set of parallel streets at a roughly perpendicular angle. (2) The only parameter is the angular separation threshold. In principle, the parameter is set at less than 10° in the regular case and is set at less than 20° in lower degradation cases, which is perceptually satisfactory.

The disadvantages and limitations are: (1) It is incapable of distinguishing between different grids. The approach proposed by Heinzle et al. (2005) can do this task; (2) The proposed approach requires that the initial street network data must be composed of straight line segments. Therefore, the data is digitized into line segments. (3) Cases similar to Fig. 10.13 will be accepted as a recognized result by the approach. In fact, these cases are bad results. Other evaluation measures to exclude these exception cases should be defined.

10.6 Conclusions and Future Work

This study is significant in that on one hand, the key questions in map generalization have been answered and on the other, the research field in pattern recognition has been extended. Inspired by the graph theory approach to the solution of object recognition problem, this paper presents a graph based approach to the recognition of grids in street networks. This approach takes the geometric features as necessary conditions. Through the graph-theoretic operations such as finding connected components, finding maximal complete subgraphs, intersections and joins, the candidates are produced. Finally, the street intersection graph is formed and the average clustering coefficient computed. The experiments illustrate that this graph based approach can be used as an effective method for recognition of grid patterns. Compared with the point-polygon based recognition approaches, the one described in this paper is a line-based recognition approach.

Some issues deserve further research. (1) The experiment was only conducted on simple data. Experiments using more complex data need to be conducted. The evaluation measures which exclude the exception case need to be developed. (2) Comparisons with other methods to solve this problem need to be made to find the best method for specific types of cases. (3) Recognition methods for other patterns in street networks need to be developed.

Acknowledgement This research was supported by the National High-Tech Research and Development Plan of China under the grant No.2009AA121404, and the China Postdoctoral Science Foundation funded project under the grant No. 20100480863. Special thanks are due to an anonymous reviewer for constructive comments which substantially improve quality of the paper. We also thank Lawrence W. Fritz for polishing up our English.

References

Batagelj V, Mrvar A (2008) Pajek: Program for analysis and visualization of large networks. Version 1.22

Diestel R (2006) Graph theory, 3rd edn. Springer, Berlin

Heinzle F, Ander KH (2007) Characterising space via pattern recognition techniques: identifying patterns in road networks. In: Mackaness WA, Ruas A, Sarjakoski LT (eds) Generalisation of geographic information: cartographic modelling and applications. Elsevier, Amsterdam, pp 233–253

Heinzle F, Ander KH, Sester M (2005) Graph based approaches for recognition of patterns and implicit information in road networks. In: Proceedings of 22nd international cartographic conference, A Coruna, Spain

Heinzle F, Ander KH, Sester M (2006) Pattern recognition in road networks on the example of circular road detection. In: Raubal M, Miller HJ, Frank AU, Goodchild MF (eds) Geographic information science, GIScience2006, vol 4197. LNCS, Munster, pp 253–267

Iglin S (2003). grTheory-graph theory toolbox, http://www.mathworks.com/matlabcentral/fileexchange/4266. Accessed 30 Sep 2009

Ip H, Wong WH (1997) Detecting perceptually parallel curves: criteria and force-friven optimization. Comput Vis Image Und 68(2):190–208

Jiang B, Claramunt C (2004) A structural approach to the model generalization of urban street network. GeoInformatica 8(2):157–173

Mackaness W, Edwards G (2002) The importance of modelling pattern and structures in automated map generalization. In: Joint ISPRS/ICA workshop multi-scale representation of spatial data, Ottawa/Canada

Mackaness WA (2007) Understanding Geographic Space. In: Mackaness WA, Ruas A, Sarjakoski LT (eds) Generalisation of Geographic Information: Cartographic Modelling and Applications, Elsevier, Amsterdam, pp 1–10

Marshall S (2005) Streets and patterns. Spon Press, New York

Porta S, Crucitti P, Latora V (2006) The network analysis of urban streets: a dual approach. Physica A 369:853–866

Sarkar S, Boyer L (1994) A computational structure for preattentive perceptual organization: graphical enumeration and voting methods. IEEE Trans Man Cybern 24(2):246–266

Theodoridis S, Koutroumbas K (2009) Pattern recognition, 4th edn. Elsevier, Amsterdam

Thomson RC, Richardson DE (1999). The "Good continuation" principle of perceptual organization applied to the generalization of road networks. In: Proceedings of 19th international cartographic conference, Ottawa

Wang JY et al (1993) Principles of cartographic generalization. Surveying and Mapping Press, Beijing

Watts DJ, Strogatz SH (1998) Collective dynamics of 'small-world' networks. Nature 393 (4):440–442

Xie F, Levinson D (2007) Measuring the structure of road networks. Geogr Anal 39(3):336–356

Yang BS, Luan XC, Li QQ (2010) An adaptive method for identifying the spatial patterns in road networks. Comput Environ Urban 34(1):40–48

Zhang QN (2004). Modeling structure and patterns in road network generalization. In: ICA workshop on generalization and multiple representation, Leicester

Chapter 11
Digital Image Processing for Spatial Object Recognition via Integration of Nonlinear Wavelet-Based Denoising and Clustering-Based Segmentation

Zhengmao Ye and Habib Mohamadian

Abstract Spatial digital image analysis plays an important role in information decision support systems, especially for regions frequently affected by hurricanes and tropical storms. For aerial and satellite imaging based pattern recognition, it is unavoidable for these images to be affected by various uncertainties, such as atmospheric medium dispersion. Image denoising is thus necessary to remove noise and retain important digital image signatures. The linear denoising approach is suitable for slow variation noise cases. However, the spatial object recognition problem is essentially nonlinear. Being a nonlinear wavelet based technique, wavelet decomposition is effective for denoising blurred spatial images. The digital image is split into four subbands, representing approximation and three details (high frequency features) in the horizontal, vertical and diagonal directions. The proposed soft thresholding wavelet decomposition is simple and efficient for noise reduction. To further identify the individual targets, a nonlinear K-means clustering based segmentation approach is proposed for image object recognition. Selected spatial images were taken across hurricane affected Louisiana areas. In addition for the evaluation of this integration approach via qualitative observation, quantitative measures are proposed on the basis of information theory. Discrete entropy, discrete energy and mutual information are applied for accurate decision support.

Z.M. Ye (✉)
Department of Electrical Engineering, Southern University, 426 PBS Pinchback Hall, Baton Rouge, LA 70813, USA
e-mail: zhengmaoye@engr.subr.edu

H. Mohamadian
College of Engineering, Southern University, Baton Rouge, LA 70813, USA
e-mail: mohamad@engr.subr.edu

A. Yeh et al. (eds.), *Advances in Spatial Data Handling and GIS*, Lecture Notes in Geoinformation and Cartography,
DOI 10.1007/978-3-642-25926-5_11,

11.1 Introduction

Spatial image processing has many potential applications in the fields of ground surveillance, weather forecasting, target detection, environmental exploration, and so on. Remotely taken images are affected by various factors, such as atmospheric dispersions and weather conditions. Thus spatial images contain diverse types of noise, both slow varying and rapid varying noise. Discrete wavelet denoising can be designed to eliminate noises presented in images so as to preserve the characteristics across all frequency ranges. This involves three steps: linear wavelet transform, nonlinear thresholding and linear inverse wavelet transform. Using discrete wavelet transform, a digital image can be decomposed into an approximation component and detailed components (horizontal, vertical, diagonal). The approximation component is further decomposed. Information loss between two successive decomposition levels of approximations is represented in detail coefficients (Gonzalez and Woods 2007; Duda et al. 2000; Haykin 1999; Ye et al. 2007a, 2008). The essence of fractal-based denoising in the wavelet domain has been used to predict the fractal code of a noiseless image. Cycle spinning is incorporated into these fractal-based methods to produce enhanced estimations of the denoised images (Ghazel et al. 2006). A new image denoising method based on Wiener filtering for soft thresholding is proposed. It shows a high and stable SNR (signal to noise ratio) gain for all noise models used. This process leads to an improvement of phase images when real and imaginary parts of wavelet packet coefficients are filtered independently (Lorenzo-Ginori and Cruz-Enriquez 2005). Two techniques for spatial video denoising using wavelet transform are utilized: discrete wavelet transform and dual-tree complex wavelet transform. An intelligent denoising system is introduced to make a tradeoff between video quality and the time required for denoising. The system is suitable for real-time applications (Mahmoud et al. 2008).

Image segmentation is a main step towards automated object recognition systems. The quality of spatial images is directly affected by atmospheric medium dispersion, pressure and temperature. It emphasizes the necessity of image segmentation, which divides an image into parts that have strong correlations with objects to reflect the actual information collected (Gonzalez and Woods 2007; Duda et al. 2000; Haykin 1999). Spatial information enhances the quality of clustering. In general, the fuzzy K-means algorithm is not used for color image segmentation and is not sufficiently robust to deal with noise. In this case, integration of discrete wavelet denoising and nonlinear K-means segmentation provides a suitable solution. Spatial information can be incorporated into the membership function for the clustering of color images. For optimal clustering, gray level images are used. The spatial function is the summation of the membership function in the neighborhood of each pixel under consideration. It yields more homogeneous outcomes with fewer noisy spots. Image segmentation refers to the process of partitioning a digital image into multiple regions. Each pixel in a region is similar with respect to a specific characteristic, such as color, brightness, intensity or texture. (Ye et al. 2005, 2006; Jaffar et al. 2009). To minimize the effects from medium dispersion,

K-means clustering is critical for image processing. It is used to accumulate pixels with similarities together to form a set of coherent image layers. For K-means clustering, optimization can be implemented via the control algorithms such as the nearest neighbor rule or winner-take-all scheme. Nonlinear K-means clustering is used in this case for image segmentation (Ye et al. 2005, 2006, 2007b; Jaffar et al. 2009; Ye 2005).

To objectively measure the impact of the technology integration of image denoising and image segmentation, metrics of discrete entropy, discrete energy, relative entropy and mutual information can be introduced to evaluate all the measurable outcomes of image processing integration (MacKay 2003).

11.2 Discrete Wavelet Transform

Two spatial source images were taken from the State of Louisiana regions, frequently affected by hurricanes. The first image shows a spatial view of New Orleans and the second image shows a spatial view of Baton Rouge. The source images are contaminated by noise. The objective is to identify the diverse types of targets involved. Image processing technology integration is proposed, where nonlinear wavelet denoising is applied and nonlinear K-means clustering is used for target identification (Figs. 11.1 and 11.2).

Discrete wavelet transform uses a set of basic functions for image decomposition. In a two dimensional case, four functions are constructed: a scaling function $\varphi(x, y)$ and three wavelet functions $\psi^H(x, y)$, $\psi^V(x, y)$ and $\psi^D(x, y)$. Four product terms produce the scaling function (11.1) and separable directional sensitive wavelet functions (11.2)–(11.4), resulting in quaternary tree structure. The scaling function and wavelet functions are all determined by the Haar Transform.

Fig. 11.1 Source spatial image of New Orleans areas

Fig. 11.2 Source spatial image of Baton Rouge areas

$$\varphi(x, y) = \varphi(x)\varphi(y) \tag{11.1}$$

$$\psi^H(x, y) = \varphi(y)\psi(x) \tag{11.2}$$

$$\psi^V(x, y) = \varphi(x)\psi(y) \tag{11.3}$$

$$\psi^D(x, y) = \psi(x)\psi(y) \tag{11.4}$$

The wavelets measure variations in three directions, where $\psi^H(x, y)$ corresponds to variations along columns (horizontal), $\psi^V(x, y)$ corresponds to variations along rows (vertical) and $\psi^D(x, y)$ corresponds to variations along diagonal directions. The scaled and translated basic functions are defined by:

$$\Phi_{i,m,n}(x, y) = 2^{j/2}\varphi(2^j x - m, 2^j y - n) \tag{11.5}$$

$$\Psi^i_{i,m,n}(x, y) = 2^{j/2}\Psi^i(2^j x - m, 2^j y - n), i = \{H, V, D\} \tag{11.6}$$

where index i identifies the directional wavelets of H, V, and D. Given the image size as M by N, the discrete wavelet transform of the function f(x, y) is formulated as:

$$w_\varphi(j_0, m, n) = \frac{1}{\sqrt{MN}} \sum_{x=0}^{M-1} \sum_{y=0}^{N-1} f(x, y)\varphi_{j_0,m,n}(x, y) \tag{11.7}$$

$$w^i_\psi(j, m, n) = \frac{1}{\sqrt{MN}} \sum_{x=0}^{M-1} \sum_{y=0}^{N-1} f(x, y)\psi^i_{j,m,n}(x, y) \tag{11.8}$$

where i = {H, V, D}, j_0 is the initial scale, the $w_j(j_0, m, n)$ coefficients define the approximation of f(x, y), $w_\psi{}^i(j, m, n)$ coefficients represent the horizontal, vertical and diagonal details for scales $j >= j_0$. Here $j_0 = 0$ and select $N + M = 2^J$ so that $j = 0, 1, 2, \ldots, J-1$ and $m, n = 0, 1, 2, \ldots, 2^j - 1$. The f(x, y) can also be obtained via the inverse discrete wavelet transform. Discrete wavelet decomposition and thresholding will both be applied in the discrete wavelet transform.

Discrete wavelet transform is implemented as a multiple level transformation, where two level transformation is implemented in context. The decomposition outputs at each level include: the approximation, horizontal detail, vertical detail and diagonal detail. Each is one quarter the size of its original image followed by downsampling by a factor of two. The approximation is further decomposed into multiple levels while the detail components will not be decomposed. Information loss between two immediate approximations is captured as detail coefficients. For denoising using discrete wavelet transforms, only wavelet coefficients of the details at level one will be subject to thresholding, while the approximation components at level one and higher levels will stay the same for image reconstruction.

In a thresholding process, the selection of the threshold is critical. Soft thresholding is selected instead of hard thresholding. This choice will shrink nonzero wavelet coefficients towards zero. Considering that, in general, a small threshold produces a good but still noisy estimation while a big threshold produces a smooth but blurred estimation, the median value stems from the absolute value of the wavelet coefficients at each selected wavelet decomposition level. The shrinkage function for soft thresholding is formulated at each decomposition level as (11.9), where THR is the median threshold value based on wavelet coefficients. x is the input signal and f(x) is the nonlinear signal after thresholding.

$$f(x) = \mathrm{sgn}(x)(|x| - THR) \tag{11.9}$$

Using wavelet denoising, two revised images are generated which represent the intrinsic geographical information for the two biggest cities of the State of Lousiana (Figs. 11.3 and 11.4). These two denoised images are further analyzed by nonlinear K-means clustering.

Fig. 11.3 Denoised image of New Orleans areas

Fig. 11.4 Denoised image of Baton Rouge areas

11.3 Nonlinear K-Means Segmentation

In image nonlinear segmentation, four clusters are proposed for partitioning. The centers of each cluster represent the mean values of all data points in that cluster. A distance metric should be determined to quantify the relative distances between objects. Both Euclidean and Mahalanobis distances are major types of distance metrics. Computation of distance metrics is based on spatial gray level histograms of the digital images. The Mahalanobis metric distance has been applied, which is formulated as (11.10), where X_A is the cluster center of any layer X_A, s is a data point, d is the Mahalanobis distance, and $K_A{}^{-1}$ is the inverse of the covariance matrix.

$$d = (s - X_A)^T K_A^{-1} (s - X_A) \tag{11.10}$$

K-means clustering assigns each object a space location, which classifies data sets through numbers of clusters. It selects four cluster centers and points cluster allocations to minimize errors. Optimal statistical algorithms are applied for classification, which are categorized as threshold based, region based, edge based or surface based. The distances between any specific data point and several cluster centers should be compared for decision making. For each individual input, winner-take-all competitive learning (11.11)–(11.12) is applied so that only one cluster center is updated. Images will thus be decomposed into four physical entities. In fact, the winner-take-all learning network classifies input vectors into one of the specified categories according to clusters detected in the training dataset. All points are eventually allocated to the closest cluster. Learning is performed in an unsupervised mode. Cluster centers have associated weights that are listed as w's. The winner is defined as the one whose cluster center is closest to the inputs. Thus, this mechanism allows for competition among all input responses, but only one output is active each time. The unit that finally wins the competition is the winner-take-all cluster, so the best cluster center is computed accordingly.

$$W_{I,J}X = MIN(W_1X), FOR\ J = 1,2,3,4; I = 1,2,3,4 \tag{11.11}$$

$$W_{I1} + W_{I2} + W_{I3} + W_{I4} = 1, FOR\ I = 1,2,3,4 \tag{11.12}$$

Assume the cluster center S wins, the weight increment of S is computed exclusively and then updated according to (11.13), where α is a small positive learning parameter. This decreases as competitive learning proceeds.

$$w_{jj} = \alpha(x_j - w_{jj}),\ for\ j = 1,2,3,4; i = 1,2,3,4 \tag{11.13}$$

The segmented outcomes of two city images can be obtained by K-means clustering with 4 clusters, respectively, representing major features such as highway, river, building and grass lawn.

11.4 Quantitative Analysis

11.4.1 Histogram and Probability Functions

For a M by N digital image, occurrence of the gray level is described by the co-occurrence matrix of relative frequencies. The occurrence probability function is then estimated from its histogram distribution.

11.4.2 Discrete Entropy Analysis

Discrete entropy is a measure of information content, which represents the average uncertainty of the information source. Discrete entropy is the summation of the products of the probability of the outcome and the logarithm of the inverse probability of the outcome, taking into account of all possible outcomes {1, 2, ..., n} as the gray level in the event $\{x_1, x_2, \ldots, x_n\}$, where p(i) is the probability at the gray level i, which contains all the histogram counts. The discrete entropy H(x) is formulated as (11.14) and all corresponding results are shown in Table 11.1.

$$H(x) = \sum_{i=1}^{k} p(i)\log_2\frac{1}{p(i)} = -\sum_{i=1}^{k} p(i)\log_2 p(i) \tag{11.14}$$

Table 11.1 Discrete entropy of images

Discrete entropy	Image A (N.O.)	Discrete entropy	Image B (B.T.R)
Source image	6.5630	Source image	6.7279
Denoised image	6.9670	Denoised image	7.2252
Cluster 1	1.4650	Cluster 1	2.2084
Cluster 2	3.1182	Cluster 2	3.1886
Cluster 3	2.9009	Cluster 3	3.0486
Cluster 4	0.5474	Cluster 4	2.0423

Table 11.2 Discrete energy of images

Discrete energy	Image A (N.O.)	Discrete energy	Image B (B.T.R)
Source image	0.0122	Source image	0.0112
Denoised image	0.0090	Denoised image	0.0077
Cluster 1	0.6705	Cluster 1	0.4802
Cluster 2	0.2684	Cluster 2	0.2609
Cluster 3	0.2788	Cluster 3	0.2938
Cluster 4	0.8965	Cluster 4	0.5411

11.4.3 Discrete Energy Analysis

The discrete energy measure indicates how the gray level elements are distributed. Its formulation is shown in (11.15), where E(x) represents the discrete energy with 256 bins and p(i) refers to the probability distribution functions at different gray levels, which contains the histogram counts. For any constant value of the gray level, the energy measure can reach its maximum value of one. A lower energy corresponds to a larger number of gray levels and a higher one corresponds to a smaller number of gray levels. The discrete energy of the source, denoised and segmented images are shown in Table 11.2.

$$E(x) = \sum_{i=1}^{k} p(i)^2 \tag{11.15}$$

11.4.4 Relative Entropy Analysis

Assuming that two discrete probability distributions of the digital images have the probability functions of p(i) and q(i), the relative entropy of p with respect to q is defined as the summation of all the possible states of the system, which is formulated as (11.16). The relative entropies of the source, denoised and segmented images are shown in Table 11.3.

Table 11.3 Relative entropy of images

Relative entropy	Source image	Denoised image A	Source image	Denoised image B
Cluster 1	0.0705	0.2944	0.0255	0.1938
Cluster 2	0.0882	0.2835	0.0294	0.1994
Cluster 3	0.0905	0.2846	0.0569	0.2665
Cluster 4	0.0512	0.2707	0.0719	0.3043
Denoised image	0.3167		0.3088	

Table 11.4 Mutual information between images

Mutual information	Source image	Denoised image A	Source image	Denoised image B
Cluster 1	5.0980	5.5020	4.5194	5.0167
Cluster 2	3.4447	3.8487	3.5392	4.0365
Cluster 3	3.6621	4.0661	3.6793	4.1765
Cluster 4	6.0156	6.4196	4.6855	5.1828
Denoised image	0.4040		0.4973	

$$d = \sum_{i=1}^{k} p(i) \log_2 \frac{p(i)}{q(i)} \tag{11.16}$$

11.4.5 Mutual Information Analysis

Also needing discussion is another metric for the mutual information I(X; Y), which is used to describe how much information one variable tells us about the other variable. The relationship is formulated as (11.17).

$$I(X;Y) = \sum_{X,Y} p_{X,Y}(X,Y) \log_2 \frac{p_{X,Y}(X,Y)}{p_X(X)p_Y(Y)} = H(X) - H(X|Y) \tag{11.17}$$

where H(X) and H(X|Y) are values of the entropy and conditional entropy; p_{XY} is the joint probability density function and p_X and p_Y are marginal probability density functions. It can be interpreted that information that Y can tell us about X is the reduction in uncertainty of X due to the existence of Y. The mutual information also represents the relative entropy between the joint distribution and product distribution. Calculated mutual information outcomes among the source, denoised and segmented images are indicated in Table 11.4.

From Table 11.1 to Table 11.2, the denoised images provide more useful information than source images and each individual image cluster provides partial information. From Table 11.1 to Table 11.4, the quantitative values separating the segmented images and the original images can be set as measures for target detection when more clusters are generated. Each cluster actually represents certain types of objects that need to be identified. This image processing integration approach has been successfully applied to spatial object recognition issues.

11.5 Conclusions

This paper has presented outcomes from the integration of image processing technologies. Image denoising can be used to maintain the energy of the images and reduce the energy of noises. Being a nonlinear approach, wavelet denoising has advantages when dealing with highly nonlinear spatial images. Using a set of wavelet bases, the wavelet coefficients can be thresholded to reduce the influence from noise. Wavelet denoising has been used to remove noise without distorting important features of the images. Image segmentation can be used to identify objects from images. It classifies each image pixel to a segment according to its similarity in specific metric distance. To reduce the blurring effects of spatial images stemming from atmospheric distortion, nonlinear region K-means clustering for image segmentation, where the competitive learning rule is applied to update clustering centers, has been applied with satisfactory results. To evaluate the roles of the wavelet denoising and nonlinear segmentation approaches, quantitative metrics have been proposed. Several information measures of discrete energy, discrete entropy, relative entropy and mutual information have been applied to indicate the effects of integrating two image processing approaches. These methodologies could be easily expanded to other image processing techniques for the different types of potential practical cases.

References

Duda et al. (2000) Pattern classification, 2nd edn. Wiley, ISBN: 978-0-471-05669-0, Hoboken, New Jersey, USA

Ghazel M, Freeman G, Vrscay E (2006) Fractal-wavelet image denoising revisited. IEEE Trans Image Process 15:9

Gonzalez R, Woods R (2007) Digital image processing, 3rd edn. Prentice Hall, ISBN-13: 9780131687288, Upper Saddle River, New Jersey, USA

Haykin S (1999) Neural networks: a comprehensive foundation, 2nd edn. Prentice Hall, ISBN-13: 9780131471399, Upper Saddle River, New Jersey, USA

Jaffar M, Naveed N, Ahmed B, Hussain A, Mirza A (2009) Fuzzy C-means clustering with spatial information for color image segmentation. In: Proceedings of the 2009 international conference on electrical engineering, Lahore, 9–11 April 2009, p 6

Lorenzo-Ginori J, Cruz-Enriquez H (2005) De-noising method in the wavelet packets domain for phase images. In: CIARP 2005, Springer-Verlag, pp 593–600

MacKay D (2003) Information theory, inference and learning algorithms. Cambridge University Press, New York City, New York 10013-2473, USA

Mahmoud R, Faheem M, Sarhan A (2008) Intelligent denoising technique for spatial video denoising for real-time applications. In: Proceedings of 2008 international conference on computer engineering & systems, Ain Shams University, Cairo, pp 407–12

Ye Z (2005) Artificial intelligence approach for biomedical sample characterization using Raman spectroscopy. IEEE Trans Autom Sci Eng 2(1):67–73

Ye Z, Ye Y, Mohamadian H, Bhattacharya P (2005) Fuzzy filtering and fuzzy K-means clustering on biomedical sample characterization. In: Proceedings of 2005 IEEE international conference on control applications, Toronto, Aug 2005, pp 90–95

Ye Z, Luo J, Bhattacharya P, Ye Y (2006) Segmentation of aerial images and satellite images using unsupervised nonlinear approach. WSEAS Trans Syst 5(2):333–339

Ye Z, Mohamadian H, Ye Y (2007) Information measures for biometric identification via 2D discrete wavelet transform. In: Proceedings of the 2007 IEEE international conference on automation science and engineering, Scottsdale, 22–25 Sept 2007, pp 835–840

Ye Z, Mohamadian H, Ye Y (2007) Discrete entropy and relative entropy study on nonlinear clustering of underwater and arial images. In: Proceedings of the 2007 IEEE international conference on control applications, Oct 2007, pp 318–323

Ye Z, Cao H, Iyengar S, Mohamadian H (2008) Medical and biometric system identification for pattern recognition and data fusion with quantitative measuring. Systems engineering approach to medical automation, Chapter Six, Artech House Publishers, pp 91–112, ISBN978-1-59693-164-0

Chapter 12
CCSSM: A Toolkit for Information Extraction from Remotely Sensed Imagery

Yong Ge, Chi Zhang, and Hexiang Ba

Abstract This paper presents a toolkit for information extraction from remotely sensed imagery. This toolkit is named Classification by Combining Spectral information and Spatial information upon Multiple-point statistics (CCSSM), which comprises the derivation of two probability fields from supervised spectral extraction classification and spatial information multiple-point simulation (MPS) and then the fusing the two. This toolkit includes three components: (a) processing input data which include the probability field from the supervised classification and a training image for MPS; (b) implementing MPS to obtain simulation result as the second; (c) fusing the above two probability fields. The performance of CCSSM for a multiple-class classification is demonstrated by a QUICKBIRD 2.5 m imagery which contains five classes.

12.1 Introduction

Incorporating spatial structure information and spatial correlation information together with spectral information is one way of improving the classification accuracy of remotely sensed imagery (Ge et al. 2008b). In past decades, much

Y. Ge (✉) • H.X. Ba
State Key Laboratory of Resources and Environmental Information System, Institute of Geographic Sciences & Natural Resources Research, Chinese Academy of Sciences, CAS A11, Datun Road, Beijing 100101, China
e-mail: gey@lreis.ac.cn

C. Zhang
State Key Laboratory of Resources and Environmental Information System, Institute of Geographic Sciences & Natural Resources Research, Chinese Academy of Sciences, CAS A11, Datun Road, Beijing 100101, China

School of the Earth Sciences and Resources of China University of Mine and Technology, Xuzhou, China

A. Yeh et al. (eds.), *Advances in Spatial Data Handling and GIS*, Lecture Notes in Geoinformation and Cartography,
DOI 10.1007/978-3-642-25926-5_12,

effort has been directed towards developing such excellent methods as contextual classification, classification using texture structural information, and classification utilizing geostatistics. Recently, Ge et al. (2008a,b) and Ge and Ba (2011) introduced multiple-point simulation (MPS) into the process of information extraction to increase the classification accuracy of remotely sensed imagery. MPS was used to characterize the structural and spatial association properties of geographical objects through a training image. MPS is one of the main applications of multiple-point statistics (Zhang et al. 2005), while single normal equation simulation (SNESIM) is one of the effective MPS algorithms . The method of integrating MPS with spectral information is called the Classification of Combining Spectral information and Spatial information upon Multiple-point simulation (CCSSM) (Ge et al. 2008b). The performance of CCSSM for the case of the two-class classification has been substantiated by experiment and accuracy assessment using a Landsat Thematic Mapper (TM) 30 m image (Ge et al. 2008a,b). This paper focuses to introduce the software toolkit of CCSSM followed by an example of a multiple-class classification from a QUICKBIRD 2.5 m imagery which contains five classes.

12.2 CCSSM

12.2.1 General Concept

CCSSM comprises the derivation of two probability fields from the supervised classification and multiple-point simulation (MPS) based on spectral and spatial information. These fields are then are fused. CCSSM uses the training image and the template information, to acquire structural information about geographical objects and compensates for the inadequacy of using only spectral information for information extraction. It not only takes spectral information of the remotely sensed imager into account, but also considers the spatial structure information and spatial correlation information of geographical objects. Therefore, CCSSM, through a training image, can effectively extract the investigated object including its distinct spatial structural characteristics, In particular it can assist the user to extract objects smaller than those implied by the spatial resolution of the imagery (Ge et al. 2008a,b; Ge and Ba 2011).

In previous studies, experiments undertaken (Ge et al. 2008a,b; Ge and Ba 2011) were only concerned with cases of two-class classification; for instance, road and non-road. In fact, the algorithm is not limited to such two-class situations. For example, assume that there are three categories with distinct structure characteristics. There are two ways of simulating these classes. (1) Firstly draw three training images which correspond to the simulations of the three categories. Secondly, arbitrarily select a category as the first category to be simulated with the corresponding training image and treat all other categories as one category. The second category is then simulated versus all other categories pooled together except for the first category. The third category can then be similarly simulated. (2) The second way is to first draw one training image for the three categories and then simulate the three categories simultaneously with that training image. In this paper, the second approach is implemented in the example below.

12.2.2 CCSSM Toolkit

The CCSSM toolkit is written in c++ with GTK. This toolkit includes three components for processing input data, intermediate data and output data as shown in Fig. 12.1. The input data includes the probability field from the supervised classification and a training image. The supervised classification can be made using a commercial software package designed to process remotely sensed data such as PCI Geomatica, EARDAS and IDRISI. In this paper, the commonly used maximum likelihood classification (MLC) is chosen as the supervised classification in CCSSM. The training image can be drawn using commercial image software such as Photoshop, and then converted into a raster text file as input data.

The second component processes the intermediate data which includes hard data and search tree data. The hard data for MPS can be obtained from the training data for MLC or the classified results from MLC. The search tree is a dynamic data structure that stores the conditional probability distribution of the training image and allows retrieval of all conditional probability distributions existing in a training image (Strebelle 2000; Ge et al. 2008b). The single normal equation simulation (SNESIM) is one of the effective algorithms of MPS. In CCSSM, SNESIM has been incorporated to extract spatial structural information. In the SNESIM algorithm, the template is an important parameter in the simulation process. It is a search window that consists of a set of ranked pixels and MPS uses it to capture patterns from the training image. The SNESIM algorithm with a multiple grid

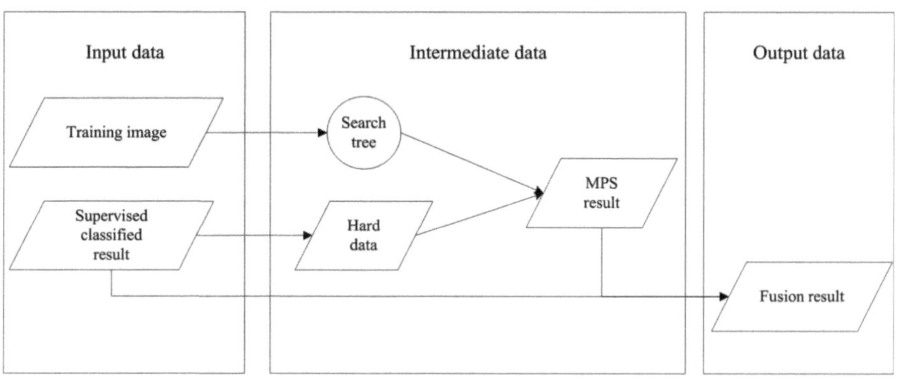

Fig. 12.1 Components and data flow in CCSSM toolkit

approach is performed to simulate the unsampled data through the training image and then to obtain the MPS result (Ge and Ba 2011).

The third component fuses the probability fields from MPS and MLC. The CCSSM toolkit provides the three fusion methods of consensus-based fusion, evidence-based fusion and probability-based fusion. The CCSSM toolkit, provides text type formats for input, intermediate and output data.

12.3 Example

The experimental procedure consists of the four steps: MLC classification, MPS, data fusion using the consensus theory, and an accuracy assessment using error matrices. As described above, the MLC classification is performed first, followed by obtaining the MPS results with the revised SNESIM algorithm, and finally fusing the results using consensus theory.

12.3.1 Data Description

The example data was selected from QUICKBIRD 2.5 m imagery. The image size is 614 $\times$ 787 pixels and its resolution is 2.5 m as shown in Fig. 12.2. From this image, it can be seen that the roads and buildings have strong spatial structural features For instance, in the study area, most roads are of a cross network structure and a certain inclination and most buildings distributed in squares in the middle of the image.

Fig. 12.2 QUICKBIRD 2.5 m imagery

Fig. 12.3 Training images (560 × 400)

12.3.2 Training Image

The training image was drawn by hand according to common conventions for objects of interest. In this remotely sensed imagery, the classes of buildings, gardens, roof shadow and streets have obvious structures and their spatial distributions are also homogeneous except for lawn areas (Fig. 12.3).

12.3.3 MLC

From Fig. 12.2, the geographical objects can be roughly classified into the 5 classes of building roof, garden between buildings, street space, lawn and roof shadow. To investigate the performance of the CCSSM method on multiple-class classification, the training data was selected according to the five classes. The MLC classification algorithm was then performed using PCI Geomatica (Fig. 12.4).

12.3.4 MPS Simulation

12.3.4.1 Hard Data

There are two ways of obtaining the hard data for MPS as shown in Fig. 12.5. One is to directly use MLC sample data as hard data. The other is to select some pixels

Fig. 12.4 MLC results

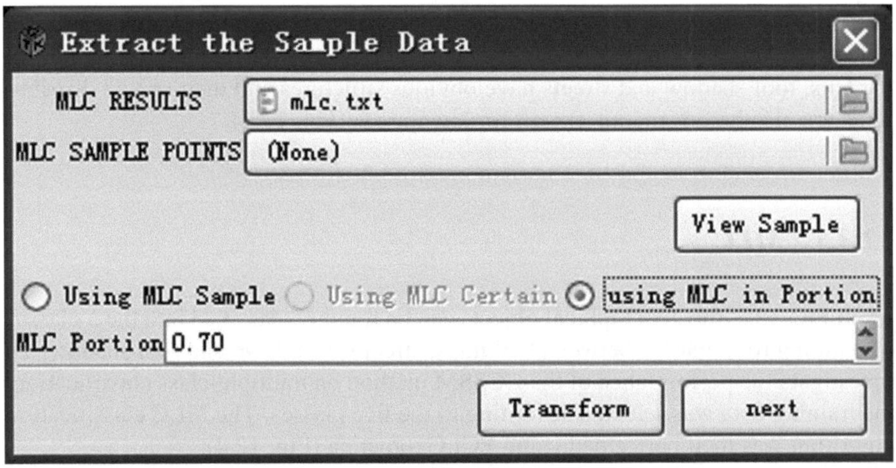

Fig. 12.5 Interface for extracting sample data for MPS

from the MLC classified results by setting a threshold such as 0.7. The threshold can be set in terms of the resolution of remotely sensed imagery and the classification accuracy of MLC.

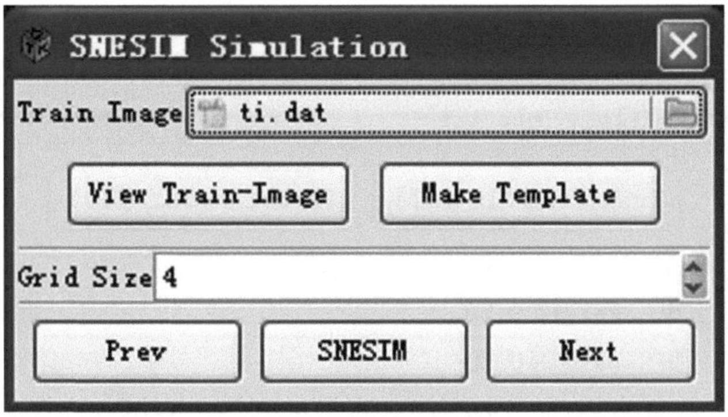

Fig.12.6 Interface for setting parameters for SNESIM

Fig. 12.7 MPS result

12.3.4.2 SNESIM

The training image is first selected and "View Train-Image" then used to view the training image as shown in Fig. 12.6. The function of the button "Make Template" is to set a template. For instance, the template can be set as 5 × 5 or7 × 7 pixels. The grid size is to set the multiple-grid approach parameter. This parameter can be set to 1, 2, 4, 8, etc.. ... "Sort Interval" selection drives the random path sorting operation after simulating a fixed number of pixels (Ge et al. 2008b). After all parameters have been set the SNESIM button is selected and the MPS result obtained as shown in Fig. 12.7.

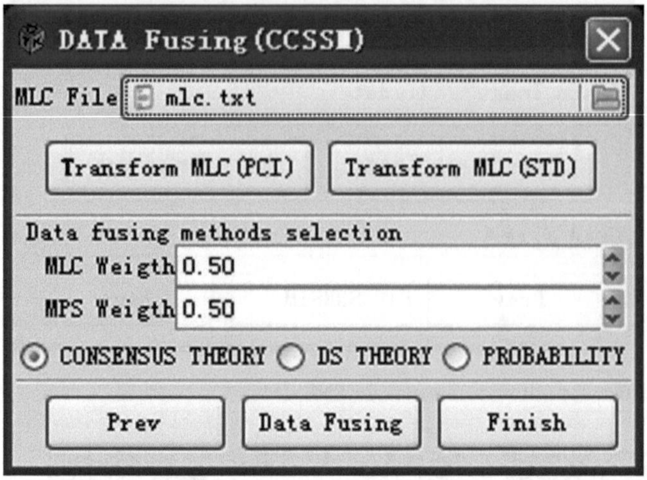

Fig. 12.8 Interface for setting parameters for fusion

Fig. 12.9 Fusion result

12.3.5 Fusion

First "Transform" button is selected to extract the probability filed from MLC and the fusion method is then selected as shown in Fig. 12.8. In this example, the CONSENSUS-based fusion method was chosen to fuse two classification results. Finally, the "Data Fusing" button is pressed to obtain the fusion result as shown Fig. 12.9. The result is then saved in text file format which can be viewed using the ENVI software package.

12.4 Conclusions and Future Works

In this paper, the general CCSSM concept and its toolkit has been introduced and its use has been demonstrated in the case of a multiple-class classification exercise. Experiments showed that the method can effectively extract information on those features of interest possessing distinct structural characteristics. Some aspects, however, need to be further discussed. The software toolkit needs further improvement to enable free downloading by interested users around the world. The template needs to be designed in a more scientific manner which better considers the spatial structure and scale characteristics of the study area. Since the training image is a key factor in MPS directly influencing the simulation results, it is essential to design a realistic training image.

Acknowledgments This work was supported in part by and the National Natural Science Foundation of China (Grant No. 40971222) and State Key Laboratory of Independent Innovation Team Project (Grant No. O88RA203SA). The authors are grateful to the editors who helped to improve the quality of the paper.

References

Ge Y, Ba HX (2011) Multiple-point simulation-based method for extraction of objects with spatial structure from remotely sensed imagery. Int J Remote Sens 32(8):2311–2335

Ge Y, Ba HX, Cheng QM (2008a) The solution of multiple-point statistics to extracting information from remotely sensed imagery. J Chinese Univ Geosci 19:421–428

Ge Y, Ba HX, Cheng QM (2008b) New classification method for remotely sensed imagery via multiple-point simulation: experiment and assessment. J Appl Remote Sens 2:023537

Strebelle S (2000) Sequential simulation drawing structures from training images. Unpublished doctoral dissertation, Stanford University, California

Zhang T, Switzer P, Journel A (2005) Merging prior structural interpretation and local data: the bayes updating of multiple-point statistics. In: Proceeding of international association for mathematical geology, 21–26 August 2005, Toronto (Toronto: the geomatics research laboratory, York University and Wuhan: China University of Geosciences), pp 615–620

Part V
Location-Based Service

Chapter 13
Evaluation of Automatically Extracted Landmarks for Future Driver Assistance Systems

Claus Brenner and Sabine Hofmann

Abstract In the future, vehicles will gather more and more spatial information about their environment, using on-board sensors such as cameras and laser scanners. Using this data, e.g. for localization, requires highly accurate maps with a higher level of detail than provided by today's maps. Producing those maps can only be realized economically if the information is obtained fully automatically. It is our goal to investigate the creation of intermediate level maps containing geo-referenced landmarks, which are suitable for the specific purpose of localization. To evaluate this approach, we acquired a dense laser scan of a 22 km scene, using a mobile mapping system. From this scan, we automatically extracted pole-like structures, such as street and traffic lights, which form our pole database. To assess the accuracy, ground truth was obtained for a selected inner-city junction by a terrestrial survey. In order to evaluate the usefulness of this database for localization purposes, we obtained a second scan, using a robotic vehicle equipped with an automotive-grade laser scanner. We extracted poles from this scan as well and employed a local pole matching algorithm to improve the vehicle's position.

13.1 Introduction

13.1.1 Driver Assistance Systems

In recent years, research and development in the area of car driver information and safety systems has been quite active. Beside the goal to provide information to the driver, active safety systems should completely prevent accidents, instead of just

C. Brenner (✉) • S. Hofmann
Institute of Cartography and Geoinformatics, Leibniz Universität Hannover, Appelstrasse 9a, Hannover 30167, Germany
e-mail: claus.brenner@ikg.uni-hannover.de; sabine.hofmann@ikg.uni-hannover.de

A. Yeh et al. (eds.), *Advances in Spatial Data Handling and GIS*, Lecture Notes in Geoinformation and Cartography,
DOI 10.1007/978-3-642-25926-5_13,

reducing the consequences. Therefore, vehicles are using more and more onboard sensors such as cameras, laser scanners or radar to gather spatial information about their environment. There are numerous examples for driver assistance systems, like adaptive cruise control, collision warning/avoidance, curve speed warning/control, and lane departure warning. Compared to car navigation, these so-called advanced driver assistance systems (ADAS) operate at a very detailed scale, requiring large scale maps, and thus, even more detailed field surveys. The NextMAP project has investigated the future map database requirements and the possible data capturing methods (Pandazis 2002). The key point is to align the evolution of in-vehicle sensors with advances in mapping accuracy and map contents in such a way that sensible ADAS applications can be realized (at minimum cost). For example, additional mapping of speed limit, stop, priority, right of way, give way and pedestrian crossing signs, traffic lights, as well as the capture of the number of lanes, the road and lane width, and of side obstacles such as houses, walls, trees, etc. has been proposed. However, the NextMAP project has not dealt with the commercial aspects of producing such maps with very high detail and accuracy. Some investigations with regard to this have been made in the 'Enhanced Digital Mapping' project (EDMap 2004). ADAS applications have been classified, according to their requirements, into 'WhatRoad', 'WhichLane', and 'WhereInLane', the latter requiring a mapping accuracy of $\pm$ 0.2 m. It was concluded that contemporary mapping techniques would be too expensive to provide such maps for the'WhichLane' and'WhereIn-Lane' case.

A key observation is that these findings assume a 'traditional' map production, i.e. the map is a highly abstract representation in terms of a vector description of the geometry, with additional attributes. However, depending on the application, this is not always necessary. For example, the accurate positioning of vehicles using relative measurements to existing map features does not necessarily require a vector map in the usual sense. Rather, any map representation is acceptable as long as it serves the purpose of accurate positioning. This means that map production to derive a highly abstract representation (which usually involves manual interaction) is not required, but rather a more low-level representation is suitable as well (which can be derived fully automatically and thus, inexpensively).

Also, accuracy considerations nowadays usually assume that the position of a vehicle relative to the map is obtained using an absolute measurement for both, typically using the global positioning system (GPS) for the vehicle and georeferenced maps (possibly obtained by GPS measurements as well). However, when the vehicle uses relative measurements to known objects in the environment (such as poles and crash barriers), a very high absolute map accuracy is not required. Thus, one is relieved from unreasonable high requirements regarding absolute map accuracy and absolute vehicle positioning, nevertheless making relative centimeter level positioning feasible.

13.1.2 Mapping Based on Mobile Laser Scanning

Mobile mapping systems using laser scanners combined with GPS and inertial sensors are well suited for the production of large scale maps, since they reach a relative accuracy of down to a few centimeters. With a rate of more than 100,000 points/s, they capture the environment in great detail (Kukko et al. 2007).

The problem of building a representation of the environment using laser scanners has also been investigated in robotics (although, of course, laser scanners were not the only sensors considered). Iconic representations, such as occupancy grids, have been used as well as symbolic representations consisting of line maps or landmark based maps (Burgard and Hebert 2008). One of the major problems to solve is the simultaneous localization and mapping (SLAM), which incrementally builds a map while a robot drives (and measures) in unknown terrain (Thrun et al. 2005). Often, 2D laser scans (parallel to the ground) are used for this, but it has also been extended to the 3D case (Borrmann et al. 2008). However, it is probable that it will still take some time until such full 3D scanners are available in vehicles. What can be expected in the near future, though, are scanners which scan in a few planes such as the close-to-production IBEO Lux scanner (IBEO 2009), which scans in four planes simultaneously.

Apparently, it would be unreasonable to provide dense georeferenced point clouds for the entire road network, as this would imply huge amounts of data to be stored and transmitted. On the other hand, as pointed out in the previous section, obtaining a highly abstracted representation usually requires manual editing, which would be too expensive. Following the work in robotics, a landmark based map would be a suitable approach. For example, to provide a map useful for accurate positioning, landmarks should fulfill the following requirements: (1) they should be unique in a certain vicinity, either by themselves or in (local) groups, (2) their position//orientation should be stable over time, (3) few of them should be needed to determine the required transformation with a certain minimum accuracy, (4) they should be reliably detectable, given the available sensors and time constraints.

In this paper we describe the use of pole-like objects as landmarks. We assess the accuracy of the extracted poles using a terrestrial survey and investigate the applicability of pole matching for the problem of vehicle localization.

13.2 Data Acquisition

In the following sections, we describe the acquisition of the two data sets used. For our experiments, we obtained two different data sets of the same area in Hannover, Germany. The first data set is a very dense point cloud, obtained by the Streetmapper mobile mapping system and is described in Sect. 2.1. The second data set was acquired by a vehicle, equipped with an automotive-grade scanner and is described in Sect. 2.2.

13.2.1 Reference Data

A dense laser scan of a number of roads in Hannover, Germany, was acquired using the Steetmapper mobile mapping system. This system was jointly developed by 3D laser mapping Ltd., UK, and IGI mbH, Germany (Kremer and Hunter 2007). The scan was acquired with a configuration of four scanners. Two scanners Riegl LMS-Q120 were pointing up and down at an angle of 20° one was pointing to the right at an angle of 45°. Another Riegl LMS-Q140 was pointing to the left at an angle of 45°. The LMSQ120 has a maximum range of 150 m and a ranging accuracy of 25 mm. All scanners were operated simultaneously at the maximum scanning angle of 80° and scanning rate of about 10,000 points/s. Positioning was accomplished using IGI's TERRAcontrol GNSS/IMU system which consists of a NovAtel GNSS receiver, IGI's IMU-II fiber optic gyro IMU operating at 256 Hz, an odometer, and a control computer which records all data on a PC card for later post processing to recover the trajectory. The scanned area contains streets in densely built-up regions as well as highway like roads. The total length of the scanning trajectory is 21.7 km, 70.7 million points were captured. On average, each road meter is covered by more than 3,200 points.

13.2.2 Vehicle Data

The second scan was obtained by RTS Hanna, a vehicle which was developed by the Real Time Systems Group (RTS) at the Institute for Systems Engineering at the Leibniz Universität Hannover, Germany (Fig. 13.1). As main sensor they use two pairs of custom-built 3D laser range scanners for environmental detection, the 'RTS-ScanDriveDuo'. Each pair consists of two 2D SICK LMS 291-S05 laser scanners, which are mounted on the ScanDrive, a rotary unit. In vertical direction, the scanners cover an area of 90°, when reading remission values. Based on the center point of a single scanner, 20° of the upper half-space and 70° of the lower half-space are recorded. The angular resolution is 1° in vertical direction. The horizontal scan range of each scanner pair is 260°, with an angular resolution of approximately 2°. Taking the driving direction as 0°, the left scanner covers the area from—210° to + 50°, the right scanner covers the area from—50° to + 210°. With a rotational speed of 150°/s, it takes 2.4 s for a full rotation. Using a pair of scanners where each scanner only covers one half-space, the scanning time is reduced to 1.2 s. New scan lines are provided every 13 ms. Therefore, each 3D scan consists of 180 scan lines. The scanners have a maximum range of 30 m and a ranging accuracy of 60 mm.

For positioning, RTS-Hanna uses a Trimble AgGPS 114 without differential corrections. A speed sensor at the gear output and an angular encoder at the steering provide the odometer data.

Fig. 13.1 The RTS Hanna robotic vehicle. The point clouds were acquired using the four laser scanners mounted on the two Scan-Drives on top of the vehicle. Image courtesy of RTS

For our experiments, all data, including GPS positions and headings, odometer data and scanned point clouds, were stored on a PC card for post processing. A single point cloud contains post processed data of one scanner rotation calculated relative to fixed center point. Each scan consists of approximately 6,000 object points.

To compare results, this second scan was acquired along the trajectory of the first scan, excluding highway like roads because of speed limitations.

13.3 Feature Extraction

We decided to use poles, such as poles of traffic signs, traffic lights, and trees, as landmarks. Poles are usually abundant in street corridors, especially in inner city scenes. It has been shown earlier by (Weiss et al. 2005), that a combination of GPS, odometry, and a (four-plane) laser scanner can be used to estimate the position of a vehicle, given a previously acquired map of poles. The reported accuracies were in the range of a few centimeters.

In order to form matches, poles have to be extracted in both the reference scan and the vehicle scan, as described in the following two subsections.

13.3.1 Extraction of Poles from Streetmapper Data

The automatic extraction of simple shapes (like cylinders) from laser scanning data has already been done in other contexts, e.g. the alignment of multiple terrestrial scans in industrial environments (Rabbani et al. 2007). However, as can be seen from Fig. 13.2, left, single poles are not hit by very many scan rays. Thus, methods which rely on the extraction of the surface, of surface normal vectors, or even of curvature, are not applicable.

Instead, we use a simple geometric model for pole extraction, assuming that poles are upright, there is a kernel region of radius r_1, where laser scan points are required to be present, and a hollow cylinder, between r_1 and r_2 ($r_1 < r_2$) where no points are allowed to be present (Fig. 13.2, right). The structure is analyzed in stacks of hollow cylinders. A pole is confirmed when a certain minimum number of stacked cylinders is found. After a pole is identified, the points in the kernel are used for a least squares estimation of the pole center. The method also extracts some tree trunks of diameter smaller than r_1, which we do not attempt to discard.

In the entire 22 km scene, a total of 2,658 poles were found fully automatically, which is one pole every 8 m on average (Figure 13.3). In terms of data reduction, this is one extracted pole per 27,000 original scan points. Although the current implementation is not optimized, processing time is uncritical and yields several poles per second on a standard PC. Note that this method owes its reliability to the availability of full 3D data, i.e., the processing of the 3D stack of cylinders. We classified the extracted poles manually and found 46% trees, 21% street lights, 5% traffic lights, 5% tram poles, 8% sign posts and other poles and 9% of non identifiable pole-like objects. The group of non identifiable objects contains all pole-like objects which are located along roads, but could not be identified definitely by manual inspection, e.g. since too few laser beams had hit the pole. The error rate (false positives) was about 6% on average, with higher rates of up to more

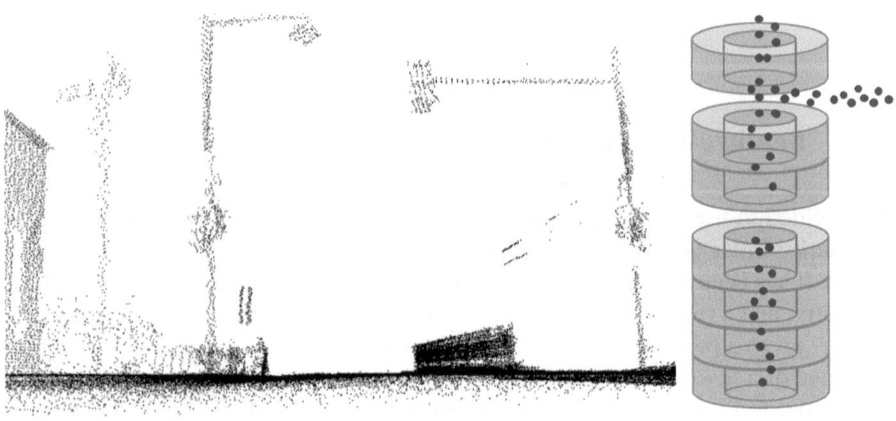

Fig. 13.2 *Left*: Part of the Streetmapper scene containing a sign post, street (and traffic) light, and traffic light. *Right*: Illustration of the pole extraction algorithm using a stack of hollow cylinders

Fig. 13.3 Extracted poles (*dots*) in a larger area with buildings from the cadastral map (*gray polygons*). The thin dotted line is the trajectory of the Streetmapper van.

Fig. 13.4 RTS Hanna data with low point density. Ground points were removed for better visibility

than 30% in densely built-up areas. False positives are mainly due to occlusions which generate narrow bands of scan columns, similar to poles.

13.3.2 Extraction of Poles from RTS Hanna Data

Hanna scan data consists of scan frames, where one frame is a horizontal sweep of any of the four scanners. One orientation is given for each single frame. As the vehicle moves during the scan, scanpoints are motion compensated in such a way that the single, given orientation holds for all points of the frame.

Unfortunately, applying the same algorithm as in the Streetmapper case to extract poles does not work well. The major problem is the low point density. From the 3D point cloud of a single scan, it is almost impossible to tell if a column of stacked points is due to an actual pole or just a consequence of the low density (Fig. 13.4). On the other hand, if one tries to combine several scans along the trajectory in order to obtain a higher density, the positioning errors will lead to several images of the very same pole, as can be seen in Fig. 13.5. The extracted poles may be more than one meter apart, even if they were derived from successive scans.

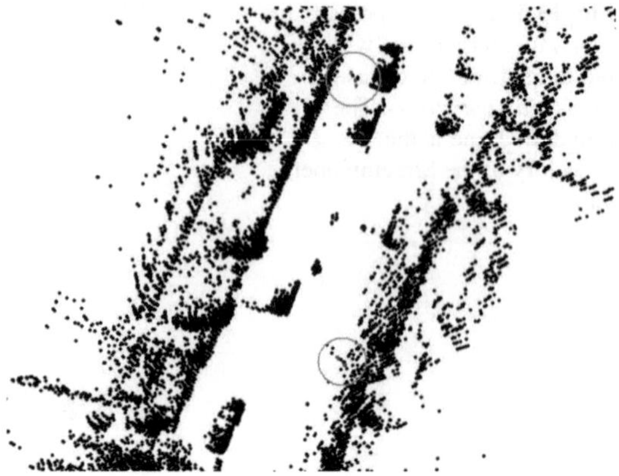

Fig. 13.5 Poles (*dots* inside *circles*) extracted from RTS Hanna data (*dots*). Ground points were removed for better visibility. Extracted poles appear at several locations due to the positioning error of the vehicle

Therefore, we extracted the poles from single scans, using the raw data rather than the 3D point cloud. Since the points are recorded in succession, column by column, a neighborhood can be defined on the row and column indices. This means that we perform our analysis on the depth image, indexed by the horizontal and vertical scan angles. One has to keep in mind, though, that this image is not an exact polar representation due to the forward movement of the vehicle during the scan. We extract poles by searching for vertical columns which have height jumps to both their left and right neighboring columns. The resulting point subsets are checked for a minimal height. Then, a principal component analysis is performed and the selected column subset is accepted based on the eigenvalues, equivalent to fitting a line and checking the residual point distances.

Figures 13.4 and 13.5 show poles which were extracted using this method. In total, 1,248 poles were detected in 1,384 scans of the first trajectory and 1,794 poles in 3,237 scans for the second trajectory. Due to the limited range and resolution of the scanners, there are around 50% (first trajectory) to 60% (second trajectory) of scans where no poles were found and around 30% of scans where only one pole was found. In both trajectories, the maximum number of detected poles in any scan was six. Figure 13.6 shows the distribution of the number of scans in which n ($0 \leq n \leq 6$) poles were found.

Figure 13.7 compares the reference poles, extracted using the Streetmapper van, with the poles obtained from the RTS Hanna scans. The latter were extracted in each scan separately, as described above. They were then transformed to the global coordinate system using the transformation provided for each scan frame, which is based on GPS positioning and odometry. In Fig. 13.7, left, the situation is especially obvious at the parking lot in the upper right, where a systematic shift is present (corresponding poles are marked by circles). Again, it can be seen that single poles appear multiple times in the RTS scans due to the low positioning accuracy. Figure 13.7, right, shows another scene at an intersection, where the reference poles are to the left and right of the street, whereas the poles extracted from RTS Hanna often appear inside the buildings or in the middle of the road. See Fig. 13.12 for final results on these two scenes.

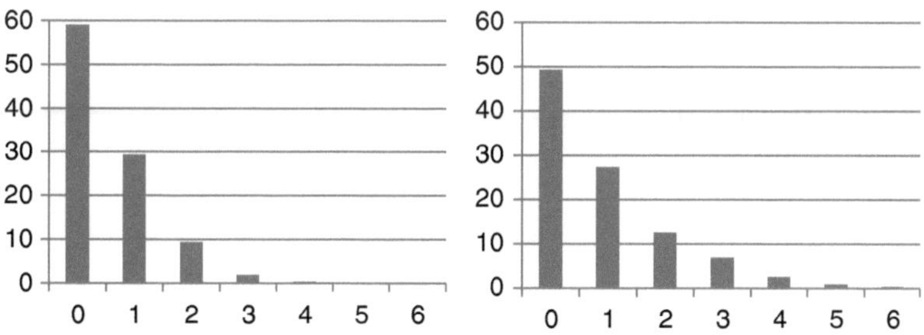

Fig. 13.6 Histograms of the percentage of scans with n extracted poles, for the first (*left*) and second (*right*) trajectory of the RTS vehicle

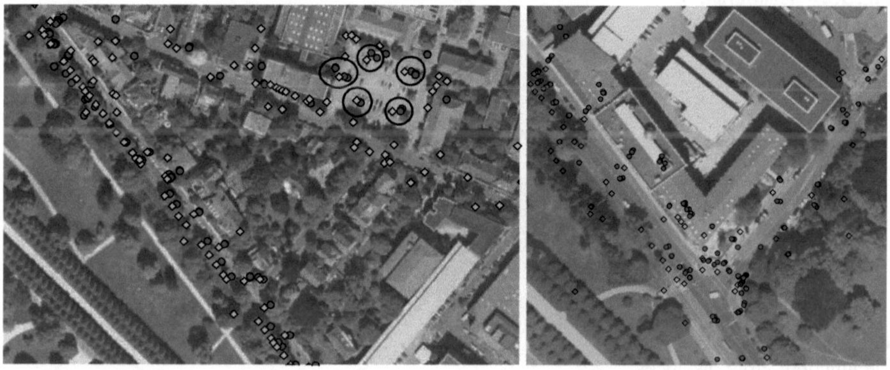

Fig. 13.7 Comparison of reference poles (*light gray squares*) and poles extracted using the RTS Hanna vehicle (*dark gray dots*)

13.4 Data Analysis

13.4.1 Verification of Reference Data

Before using the reference data provided by the Streetmapper system, we have to make a statement on the quality of this data. Therefore, an analysis of the accuracy of position of the extracted landmarks is required. Ground truth was obtained along a selected inner-city junction by a terrestrial survey, using a total station to measure every single pole manually. The accuracy of position is <10 cm for every measured pole.

As ground truth and extracted reference poles are both given in DHDN 1990, third meridian strip, calculating the residuals is very easy. We found 28 poles with a maximum distance of 32 cm from their corresponding pole. Using larger search radii only leads to mismatches between poles. Calculating the residuals using these points, we achieved a maximum distance between corresponding poles of 22.9 cm

Fig. 13.8 Accuracy of position: ground truth obtained by a terrestrial survey (*crosshairs*) compared to automatically extracted poles (*dots*) of the reference map

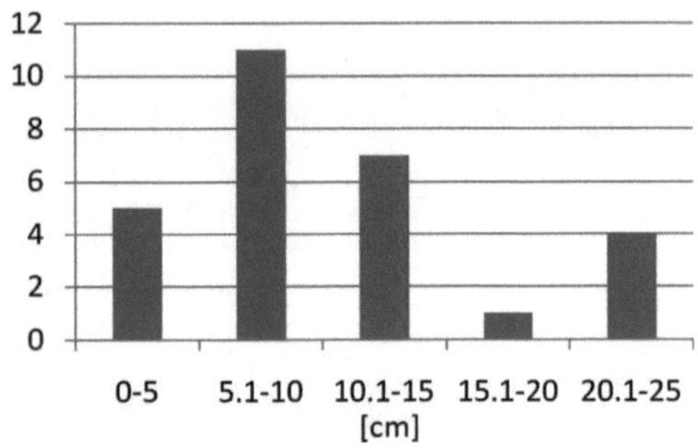

Fig. 13.9 Histogram of pole measurement errors

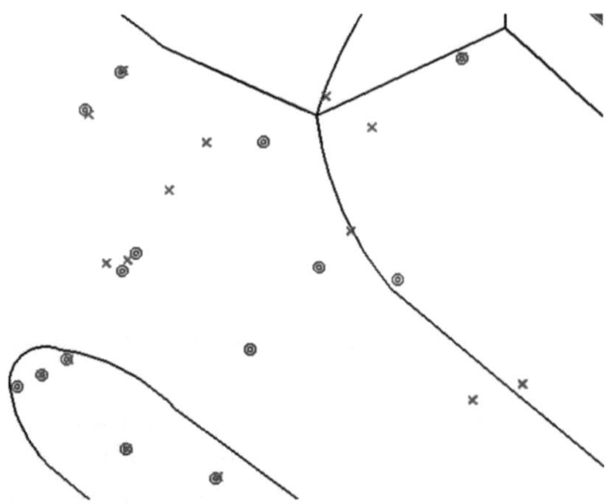

Fig. 13.10 Detailed view of the junction: large distances between extracted poles (*crosses*) and ground truth (*circles*, *radius*: *inner circle* 25 cm, *outer circle* 50 cm) are not measurement errors but due to a rebuilt of the junction

and a minimum distance of 1.4 cm. The root mean square error was 12.1 cm. Figure 13.9 shows the error distribution.

Comparing the extracted poles to the ground truth also revealed the problem of up-to-dateness. As shown in Fig. 13.10, some of the extracted poles, marked with crosses, have a large distance to ground truth poles, marked with circles (the outer circle has a distance of 50 cm to the center of the pole). In this case there is no gross error in the feature extraction algorithm, but the junction was rebuilt between scanning the area and measuring poles manually.

13.5 Experiments

The goal of our experiment is to find out if the position of the vehicle can be determined using a matching of poles. To this end, we implemented a local matching algorithm, as described in the following (see also Fig. 13.11).

The vehicle maintains a current transformation, T, and a current set of active poles, P, which is a set of 2D points. As the vehicle moves forward, new scans become available and some of them lead to newly detected poles. Those poles are then added to the active set. At the same time, old poles above a given horizon distance ε_h are removed from the active set. The reason for this is that the positions in the active set can only be assumed to be approximately correct within a certain distance, as their transformation is based on the given scan frame orientation, which uses GPS and odometry.

Whenever the active set is updated, a match to the global database of poles, Q, is attempted. For each point in the active set, all neighbors in the database within a distance smaller than a search distance ε_s are retrieved. For any translation vector between a pole in the active set and a corresponding element among its neighbors, the active set is transformed and the number of 'hits' of all poles in the active set to the database are counted. A hit is assumed when the distance is below a 'matching' threshold ε_m. This is similar to random sampling consensus (RANSAC), however, due to the low number of elements we can afford checking all possible translation vectors instead of having to pick them randomly. The transformation T' with the largest consensus set is estimated using a least squares estimation of a similarity transform. It is then taken as the new local transformation T. We used $\varepsilon_h = 100$ m for the horizon, $\varepsilon_s = 15$ m as search radius, and $\varepsilon_m = 2$ m as match radius in our experiments.

The results for the two scenes of Fig. 13.7 can be seen in Fig. 13.12, which shows the positions of the vehicle according to the given orientation per scan frame

---

Q is the reference point set
Initialize the transformation T to identity
Initialize the current active set of points P to be empty
while new points p become available **do**
    $P \leftarrow P \cup \{p\}$
    Remove all points p' from P with $\mathrm{dist}(p',p) > \varepsilon_h$
    For each point $p \in P$ compute $N_\varepsilon(p) \leftarrow \{n | \mathrm{dist}(n,p) < \varepsilon_s\}$
    for all pairs p, q with $p \in P$, $q \in N_\varepsilon(p)$ **do**
        Compute transformation T' mapping p to q
        Map all $p \in P$ using $p = T'(p)$ and count all p for which $\exists n \in N_\varepsilon(p)$ with $\mathrm{dist}(n, p') < \varepsilon_m$
    Let T' be the transformation with the highest count
    Find closest point pairs using T' and determine $T \leftarrow$ best L.S. estimation based on those pairs

---

Fig. 13.11 Algorithm used for local point pattern matching

Fig. 13.12 Comparison of original vehicle trajectories (*dark-gray dots*), as obtained from the RTS Hanna GPS/odometry positioning system and corrected trajectories (*light-gray squares*), using pole matching

(dark-gray dots) and the corrected position after applying the transformation T (light-gray squares). For example, in the left image, the trajectory starts at the parking lot, where both positions are identical in the beginning (only the dark-gray dots are visible). After a few meters, the poles are matched and the light-gray trajectory is corrected to its true position. The corrected trajectory is also clearly better in the left and lower part, where the original trajectory obviously suffers from GPS shadowing by the trees. Similarly, in the right image, the original coordinates are far off the correct lane not only after the U-turn, but also when driving straight on (presumably also due to shadowing). As there is no ground truth for the real trajectory, e.g. obtained with a GPS/IMU system with substantially higher accuracy, our analysis does not include an evaluation for the entire trajectory. Clearly, there are also many parts in the trajectory where our local matching algorithm failed so far. Sometimes, too few poles lead to failure. On the other hand, along the alleys, there are often too many poles and our local matching algorithm finds wrong correspondences, leading to a wrong transformation.

13.6 Discussion and Outlook

In this paper, we presented an approach to use automatically extracted landmarks for the positioning of vehicles. We focused on pole-like objects and extracted them both in a high-resolution, high accuracy scan (the reference database) as well as in low-resolution, low accuracy scans of a robotic vehicle, using two different algorithms. We assessed the accuracy of the extracted poles and made first experiments regarding the improvement of the trajectory using a local pole matching.

As we mentioned, pole matching did not succeed in all cases. Too few as well as too many poles may cause problems, and erroneous poles in the vehicle scans and in the database lead to wrong feature associations and thus, transformations. We think that a key to a successful matching is the short-term stability of the vehicle trajectory. If a high precision of the poles can be obtained locally, many false matches can be ruled out by tight distance bounds. However, even then, there is sometimes the problem of too few poles in the database and it is obvious that a key element for a reliable localization will be to include additional landmarks.

For the future, we plan to include such additional features, e.g. local planes. Also, the local matching algorithm should be extended and all measurements should be included in a Kalman filter solution to estimate the vehicle position.

Acknowledgements We thank Matthias Hentschel from the Real Time Systems Group at the Institute for Systems Engineering of the Leibniz Universität Hannover for providing the RTS Hanna scans.

References

Borrmann D, Elseberg J, Lingemann K, Nüchter A, Hertzberg J (2008) Globally consistent 3D mapping with scan matching. J Robot Auton Syst 56(2):130–142

Burgard W, Hebert M (2008) World modeling. In: Siciliano B, Khatib O (eds) Springer handbook of robotics. Springer, Heidelberg, Germany, pp 853–869

EDMap (2004) Enhanced digital mapping project final report. Technical report, United States department of transportation, federal highway administration and national highway traffic and safety administration. http://ntl.bts.gov/lib/jpodocs/repts_te/14161.htm. Accessed 14 July 2011

IBEO (2009) Ibeo Lux laser scanner. http://www.ibeo-as.com. Accessed 14 July 2011

Kremer J, Hunter G (2007) Performance of the streetmapper mobile lidar mapping system in real world projects. In: Fritsch D (Ed.) Photogrammetric Week 2007. Wichmann, Heidelberg, pp 215–225

Kukko A, Andrei C-O, Salminen V-M, Kaartinen H, Chen Y, Rönnholm P, Hyyppä H, Hyyppä J, Chen R, Haggrén H, Kosonen I, Capek K (2007) Road environment mapping system of the finnish geodetic institute – FGI roamer. In: IAPRS (ed.), Proceedings of laser scanning 2007 and silviLaser 2007, Vol. 36 Part 3/W 52, pp 241–247

Pandazis J-C (2002) NextMAP: investigating the future of digital map databases. In: e-Safety congress, Lyon. http://www.ertico.com/nextmap-website. Accessed 14 July 2011

Rabbani T, Dijkman S, van den Heuvel F, Vosselman G (2007) An integrated approach for modeling and global registration of point clouds. ISPRS J Photogramm Remote Sens 61 (6):355–370

Thrun S, Burgard W, Fox D (2005) Probabilistic robotics. The MIT Press, Cambridge, MA

Weiss T, Kaempchen N, Dietmayer K (2005) Precise ego localization in urban areas using laserscanner and high accuracy feature maps. In: Proceedings of 2005 IEEE intelligent vehicles symposium, Las Vegas, USA, pp 284–289

Chapter 14
Temporally Adaptive A* Algorithm on Time-Dependent Transportation Network

Nianbo Zheng and Feng Lu

Abstract Traditional solutions to shortest path problems on time-varying transportation networks use traffic information only at precise moments regardless of considering the fact that the travel time through any link is dependent on the time entering that link. In this study, travel speed rather than travel time on each link is used as the time period dependent parameter to model time-dependent transportation networks, and a First-In-First-Out (FIFO) condition satisfied computational function of link travel time is then deduced based on kinematics. Finally, a temporally adaptive A* shortest path algorithm on this FIFO network is presented, where the time factor is introduced into the evaluation function, and the Euclidean distance divided by the maximum possible travel speed is used as a heuristic evaluator. An experiment on a real road network shows that the proposed algorithm is capable of foreseeing and bypassing forthcoming traffic congestion, at a cost only about 10% more in computational time than the traditional algorithm. In addition, frequent path reoptimization required with use of the traditional algorithm is effectively avoided.

N.B. Zheng (✉)
Institute of Geographic Sciences and Natural Resources Research, Chinese Academy of Sciences, 11A Datun Road, Chaoyang District, Beijing 100101, China

Beijing Institute of Satellite Information Engineering, 82 Zhichun Road, Haidian District, Beijing 100086, China
e-mail: zhengnb@lreis.ac.cn; zhengnb7811@163.com

F. Lu
Beijing Institute of Satellite Information Engineering, 82 Zhichun Road, Haidian District, Beijing 100086, China

A. Yeh et al. (eds.), *Advances in Spatial Data Handling and GIS*, Lecture Notes in Geoinformation and Cartography, DOI 10.1007/978-3-642-25926-5_14,

14.1 Introduction

It is increasingly necessary for vehicle navigation systems and map search websites to calculate fastest paths by using real-time, historical, or predicted traffic information (Fawcett and Robinson 2000; Yamane et al. 2004; Ishikawa 2005). Traditional solutions to this problem only consider traffic information at particular moments and calculate optimal paths by either classic or heuristic shortest path algorithms such as Dijkstra algorithm, A* algorithm, branch pruning algorithm, hierarchical search algorithm (Lu and Guan 2004; Fu et al. 2006; Klunder and Post 2006; Cho and Lan 2009). As a result, during travel the planned path has to be frequently re-optimized to respond to the periodic renewal of near-real-time traffic information (e.g., every 5 min). Obviously, this process is time-consuming, even if some practical accelerating techniques, such as the dynamic window scheme (Kim and Jung 2002) and the incremental search approach (Huang et al. 2007), are utilised. In any case these algorithms are incapable of computing the best path and evaluating travel time ahead from an overall perspective. This is because they do not take account of the fact that travel time through a link is dependent on the time of entering the link. Therefore, time-dependent shortest path algorithms are needed for dynamic route planning.

Time-dependent shortest path algorithms model travel time through a link as an arrival-time dependent variable, and can be resolved by such modified label-setting algorithms as Dijkstra, A*, etc., only if the FIFO condition is satisfied (Kaufman and Smith 1993; Sung et al. 2000; Horn 2000; Chabini and Lan 2002; Kanoulas et al. 2006). In practice, the travel time through a link is usually taken as time-interval dependent which ignores the fact that the travel speed may vary for different time intervals during a trip along the link, and that the FIFO condition may be violated. In view of this, Sung et al. (2000) proposed a flow speed model (FSM) which defined travel speed rather than travel time on a link as time-interval dependent variable, and gave an FIFO-satisfied time-dependent link-travel-time function. This paper elaborates the computational process of link travel time based on time-interval dependent speeds, and then uses it for the temporal adaption of the heuristic A* shortest path algorithm.

The rest of this paper is organized as follows. Section 14.2 defines a time-dependent network with time-interval dependent link travel speeds. Section 14.3 derives a computational function for an FIFO-satisfied link travel time. Section 14.4 presents a temporally adaptive A* algorithm. Section 14.5 tests the proposed algorithm on a real road network. Conclusions are drawn in Sect. 14.6.

14.2 Time-Dependent Network Model

Suppose the time T is segmented into the following intervals: $[t_0, t_1)$, $[t_1, t_2)$, ..., $[t_k, t_{k+1})$ $(k = 0, 1, \ldots, m-1, t_k < t_{k+1})$, then the time-dependent network is defined as $G = (N, A, L, V, W)$, where $N = \{0, 1, \ldots, n-1\}$ denotes the node set; $A \subseteq \{(i, j) | (i, j) \in N \times N\}$ denotes the directed link set; $L = \{l_{ij} | (i, j) \in A, l_{ij} > 0\}$ denotes the link length set; $V = \{f_{ij}(t) | (i, j) \in A, t \in T\}$ denotes the set of the time-interval dependent link travel speeds, and for each $t \in [t_k, t_{k+1}), f_{ij}(t) = v_{(i,j)k}$; $W = \{w_{ij}(t)|(i,j) \in A, t \in T\}$ denotes the set of the time-dependent link travel times, where $w_{ij}(t)$ denotes the travel time along link (i, j) when departing from node i at time t.

Let $T_{ij}(t)$ denote the arrival time at node j with departure time t from node i along link (i, j), given by $T_{ij}(t) = t + w_{ij}(t)$. Let $p = \{i_1, i_2, \ldots, i_u\}$ denote a path from node i_1 to node i_u, then the path arrival time function of path p is given by the composition of the link arrival time functions along p:

$$T_p(t) = T_{i_{u-1}i_u}(T_{i_{u-2}i_{u-1}}(T_{i_{u-2}i_{u-3}} \cdots T_{i_1 i_2}(t))), \tag{14.1}$$

Let $P(o, d)$ denote the path set from origin node o to destination node d, $ET_{od}(t)$ denote the earliest arrival time while leaving from origin node o at time t to destination node d, then:

$$ET_{od}(t) = \min\{T_p(t) : p \in P(o,d)\}, \tag{14.2}$$

Obviously, the time-dependent shortest path problem is just a problem of computing the earliest arrival time $ET_{od}(t)$.

(FIFO conditon) For each link $(i, j) \in A$, if the following inequality is satisfied, the link is FIFO-satisfied:

$$T_{ij}(s) \leq Tij(t)\forall s \leq t \tag{14.3}$$

If every link of a time-dependent network is FIFO satisfied, the network is then a FIFO network. Kaufman and Smith (1993) proved that the shortest path problem on an FIFO network can be satisfactorily dealt with by the traditional labelling algorithms.

14.3 Computation of Link Travel Time

Consider a link $(i, j) \in A$ of length l (for simplicity, the subscript (i, j) is dropped in this section) with travel speed v_k changing with the time intervals $[t_k, t_{k+1})$ ($k = 0, 1, \ldots, m–1$). If a vehicle sets off from node i at time t and reaches node j at time $T(t)$, its travel time along this link is then $T(t) - t$. On the basis of kinematics, the computational procedure for $T(t)$ is as follows:

1. If $l_k - v_k \times (t_{k+1} - t) < 0$ (see Fig. 14.1(1)), then

   $T(t) = t + l_k/v_k,$

else

2. If $l_{k+1} - v_{k+1} \times (t_{k+2} - t_{k+1}) < 0$ (see Fig. 14.1(2)), then

   $T(t) = t_{k+1} + l_{k+1}/v_{k+1},$

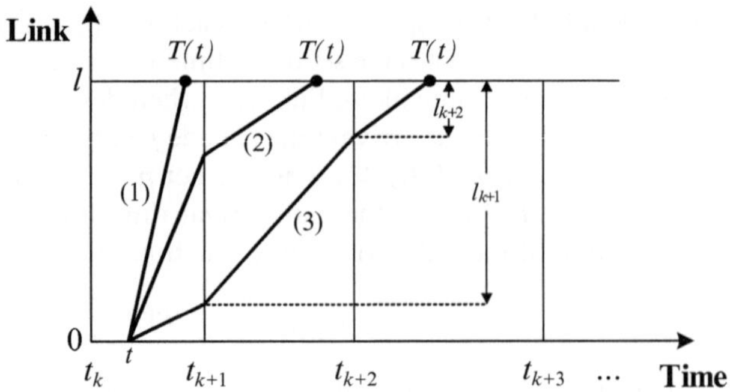

Fig. 14.1 A trip along a link with time-interval-varying speeds

else

3. If $l_{k+2} - v_{k+2} \times (t_{k+3} - t_{k+2}) < 0$ (see Fig. 14.1(3)), then

$$T(t) = t_{k+2} + l_{k+2}/v_{k+2},$$

else

$\vdots$

(i) If $l_{k+i-1} - v_{k+i-1} \times (t_{k+i} - t_{k+i-1}) < 0$, then

$$T(t) = t_{k+i-1} + l_{k+i-1}/v_{k+i-1} \forall i = 2, 3...,$$

else

$\vdots$

where

$$l_k = l$$

$$l_{k+1} = l_k - v_k \times (t_{k+1} - t)$$

$$l_{k+2} = l_{k+1} - v_{k+1} \times (t_{k+2} - t_{k+1})$$

$$\vdots$$

$$l_{k+i} = l_{k+i-1} - v_{k+i-1} \times (t_{k+i} - t_{k+i-1})$$

$$\vdots$$

In practice, if the time intervals are long enough, travel along the link will cover no more than two time intervals, namely, $[t_k, t_{k+1})$, $[t_{k+1}, t_{k+2})$. In this case, the above computational procedure can be reduced to:

$$T(t) = \begin{cases} t + l/v_k, \ t \in [t_k, t_{k+1} - l/v_k) \\ t_{k+1} + [l - v_k(t_{k+1} - t)]/v_{k+1}, t \in [t_{k+1} - l/v_k, t_{k+1}) \end{cases} \tag{14.4}$$

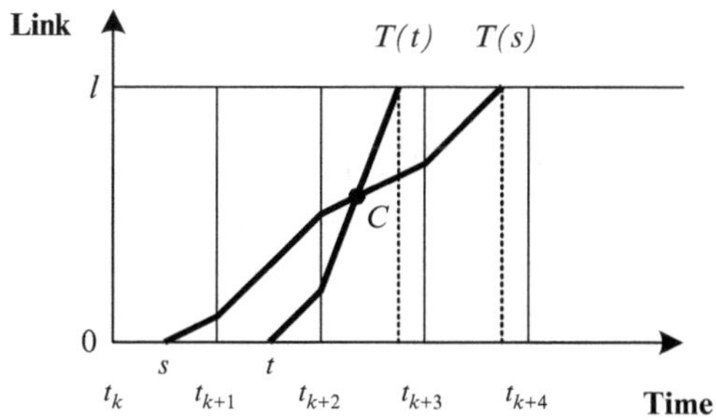

Fig. 14.2 FIFO condition violated

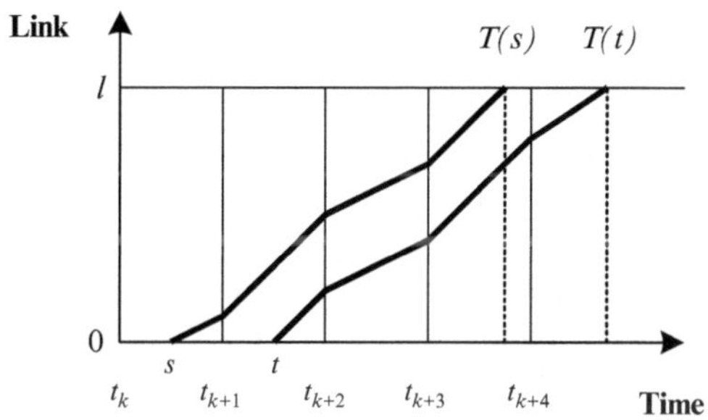

Fig. 14.3 FIFO condition satisfied

The FIFO satisficing of $T(t)$ is illustrated by Figs. 14.2 and 14.3. Assume $T(t)$ violates the FIFO condition, that is the occurrence of an overtaking then $T(s) > T(t)$ $\forall s \leq t$, and the two trajectories are sure to intersect at some position, e.g., point C in Fig. 14.2. Thus, there must exist two different speeds during the same time interval, e.g., $[t_{k+2}, t_{k+3})$ in Fig. 14.2. In the following two figures, speed is represented by the slope of the trajectory. Apparently, this violates the definition of the time-dependent network in Sect. 14.2. Consequently, it can be surely concluded that $T(t)$ satisfies the FIFO condition. In fact, two travel trajectories departing at different times will be parallel in the FIFO network, as shown in Fig. 14.3. See Sung et al. (2000) for detailed mathematical justification on this problem.

14.4 Temporally Adaptive A* Algorithm

The A* shortest path algorithm was first proposed by Hart et al. (1968), and further proved applicable to road networks by Fu et al. (2006) and Zeng and Church (2009). The A* algorithm makes use of a heuristic evaluation function $F_i = L_i + e_{(i,d)}$ as a label for node i, where L_i is the travel time of the currently evaluated path from the origin node to node i; $e_{(i,d)}$ is the estimated travel time from node i to node d. The sum of these two functions, F_i, is the weight of node i, and reflects the likelihood of node i being on the shortest path.

As for the time-dependent network, it is necessary to introduce a time factor into the evaluation function: $F_i(t) = T_i(t) + e_{(i,d)}(T_i(t))$, where t denotes the departure time from the origin node; $T_i(t)$ denotes the arrival time of the current path from the origin node at the time t to the node i; $e_{(i,d)}(T_i(t))$ is an evaluated travel time from the node i at time $T_i(t)$ to the destination node d. $e_{(i,d)}(T_i(t))$ controls the accuracy and efficiency of the algorithm. Chabini and Lan (2002) proved that if $e_{(i,d)}(T_i(t))$ is not more than the real minimum travel time from node i to the destination node d with departure time t, the temporally adaptive A* algorithm will be strictly optimal. In view of this, let

$$e_{(i,d)}(T_i(t)) = \frac{D_{(i,d)}}{V_{\max}}, \tag{14.5}$$

where $D_{(i,d)}$ denotes the Euclidean distance from node i to the destination node d; $V_{\max}$ denotes the maximum possible travel speed.

Let o denote the origin node, d the destination node, t the departure time, P_i the immediately preceding node to node i along the shortest path, Q the eligible node set for scanning purposes, and R the permanently labelled node set, then the basic steps taken by the temporally adaptive A* (TAA*) algorithm are as follows:

Step 1: Initialization: Set $T_o = t$; $F_o = T_o + e_{(o,d)}(T_o)$; $F_j = T_j = \infty \quad \forall j \neq o$; $P_o = -1$; $Q = \{o\}$; $R = \emptyset$;

Step 2: Node Selection: Select and remove the node i with the minimum label $F_i(t)$ from the scan eligible node set Q, and label it permanently:

$i = \text{argmin}_{j \in Q}\{F_j\}$; $Q = Q \backslash \{i\}$; $R = R \cap \{i\}$;

if $i == d$, the shortest path has been found, then

STOP;

else

Step 3: Node Expansion: Scan each link outgoing from node i. For each link $(i, j) \in A$ & $j \notin R$, if

$\text{Arri\_time}((i, j), T_i) + e_{(j,d)}(\text{Arri\_time}((i, j), T_i)) < F_j$,

then

$T_j = \text{Arri\_time}((i, j), T_i)$; $F_j = T_j + e_{(j,d)}(T_j)$; $P_j = i$;

insert node j into Q: $Q = Q \cap \{j\}$;

where $\text{Arri\_time}((i, j), T_i)$ is a procedure for computing the arrival time at link (i, j) according to Sect. 14.3;

Step 4: Termination: If $Q = \emptyset$, then STOP;

else: goto step 2.

14.5 Experiment

The proposed algorithm, TAA* algorithm, was tested on the already developed Urban Public Travel Path Service System with following runtime environment: dual-core CPU 1.6 GHz, RAM 1.0 GB, OS Windows XP professional.

14.5.1 Data Preparation

The experimental data include road network data within the Beijing's Fourth Ring and carriageway-based floating car traffic data of July, August, and September, 2007.

The topology of the road network is carriageway-based, and one carriageway corresponds to one link of the time-dependent network in Sect. 14.2. After generalizing those virtual links and virtual nodes that represent same road intersection into one topological node, the network contains 53,997 links in total.

One day of traffic data was split into 288 time intervals each of 5 min duration. Traffic data items in each time interval include: carriageway ID, carriageway length, travel time, and congestion level. In order to support the TAA* algorithm, these raw travel time data were converted into travel speed data by dividing carriageway length by travel time. In view of the cyclic nature of urban traffic flow, those speed values of the same day category and the same time interval were averaged, which resulted in seven speed data files.

Default speeds were provided according to road type and time of entering, as shown in Table 14.1, for the carriageways not covered by floating car traffic data, A simple time-dependent turn delay model was also built as $Turn = 0.5 \times f(time, turn\_type)$, where 0.5 denotes the probability of waiting at intersections; $f(time, turn\_type)$ is evaluated as Table 14.2.

Table 14.1 Default speed values of different road types (Speed: km/h)

	Expressway	Arterial	Sub-arterial	Minor	Overpass	Ramp
Monday morning peak (7:00–9:00) and Friday evening peak (17:00–19:00)	35	35	35	28	35	30
Other peaks (7:00–9:00; 17:00–19:00)	45	40	45	40	30	30
Nights (21:00–24:00; 0:00–6:00)	60	60	60	50	45	45
Others (6:00–7:00; 9:00–17:00; 19:00–21:00)	55	55	55	50	55	45

Table 14.2 A naive model for time-dependent turn delays (time: min)

	Straight	Right	Left	U-turn	Others
Peaks (7:00–9:00; 17:00–19:00)	1.0	0.5	1.0	0.8	0.3
Night (21:00–24:00; 0:00–6:00)	0.5	0.0	1.0	0.5	0.5
Others (6:00–7:00; 9:00–17:00; 19:00–21:00)	0.5	0.1	1.0	0.5	1.0

14.5.2 Algorithm Preparation

Three algorithms, TAA* algorithm, real-time A* (RTA*) algorithm, and multiple real-time A* (RTA*_M) algorithm, were implemented in the experiment. They all used quad-heap priority queue (Lu et al. 1999) to manage the scan eligible node set, set the maximum possible travel speed to 60 km/h, and took arc-labeling strategy (Gao and Lu 2008) to deal with turn delays. The arc-labeling strategy means that nodes and links exchange their roles and turn delays are added to the path arrival time.

The RTA* algorithm considers only those traffic data at departure time, which is significantly different from the TAA* algorithm. The RTA*_M algorithm refers to the multiple invoking of the RTA* algorithm during the overall process of vehicle navigation. It works as follows: (1) at the beginning of the trip, an optimal path is computed by the RTA* algorithm; (2) during the trip, the path is reoptimized by the RTA* algorithm as long as new up to date traffic information arrives (This was simulated using the historical traffic data set); (3) step (2) is repeated until the destination is reached. Thus the computational time exhausted by the RTA*_M algorithm is the sum of each RTA* algorithm run, and the path of the RTA*_M algorithm is that actually traveled.

14.5.3 Experimental Results

Beichendong Road was chosen as an origin road and Cuiwei Road as the destination with 7:00 and 8:00 on a Tuesday set as two departure times. Four least-time paths were then computed by the TAA* algorithm and the RTA* algorithm at departure times 7:00 and 8:00. Computational results are shown in Figs. 14.4 and 14.5.

As seen from Figs. 14.4 and 14.5, when departing at 7:00, the TAA* algorithm foresees and bypasses the forthcoming traffic congestion on the Third Ring Road, but the RTA* algorithm does not. When departing at 8:00, the current traffic congestion on the Third Ring Road will die away by the time that location is reached and therefore avoid an unwanted detour that the RTA* algorithm would postulate.

Thirty pairs of Origin–destination (O–D) carriageways, set for 7:30 on Tuesday departure times were then selected, and the least time paths for each O-D pair were calculated separately by the TAA* algorithm, the RTA* algorithm and the

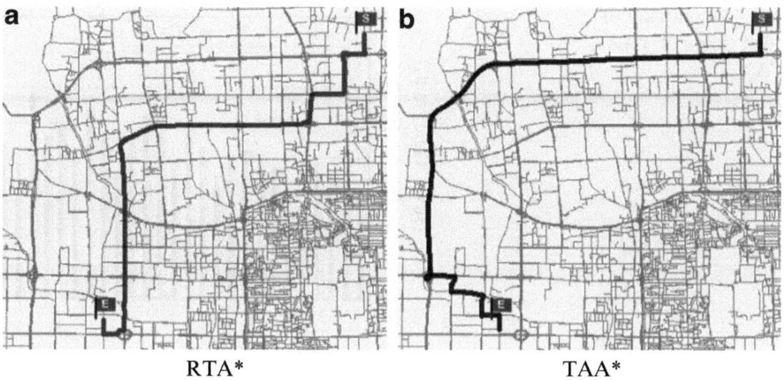

Fig. 14.4 Comparison of computational results of TAA* and RTA* (Departure time: 7:00)

Fig. 14.5 Comparison of computational results of TAA* and RTA* (Departure time: 8:00)

RTA*_M algorithm. Comparisons between them are shown in Figs. 14.6–14.8 (The 30 paths are sorted by length).

As seen from Fig. 14.6, although the low level of traffic information coverage and the immaturity of the traffic prediction model and the turn delay model make a few paths less differentiated, the path travel times calculated by the TAA* algorithm are less than those of the RTA*_M algorithm on the whole.

It is concluded from Fig. 14.7 that the TAA* algorithm will cost about 10% more in computational time than the RTA* algorithm. The extra time is taken up mainly by the link travel time computation procedure. Its time complexity is O(m), where m denotes the number of the time intervals used on the link.

It can be seen from Fig. 14.8 that the TAA* algorithm costs much less in computational time than the RTA*_M algorithm. The reason is that the TAA* algorithm only needs to perform the computational procedure once because it has

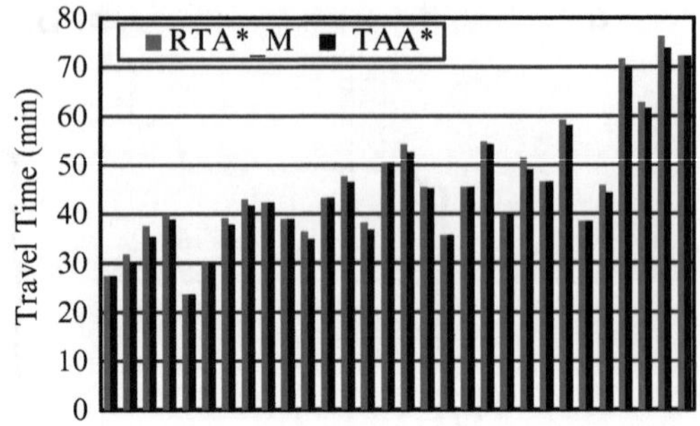

Fig. 14.6 Comparison of path travel times of TAA* and RTA*_M

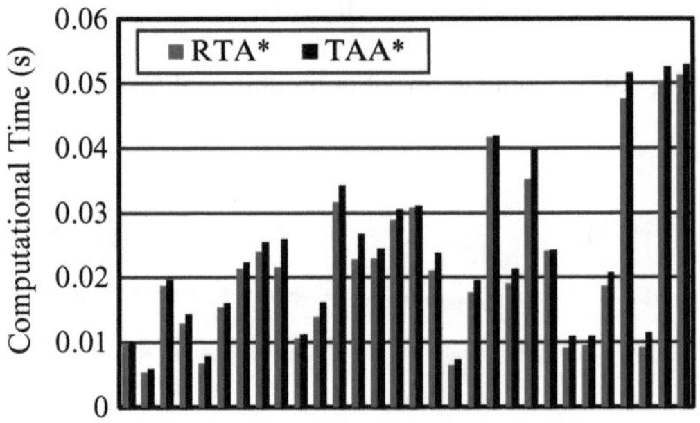

Fig. 14.7 Comparison of computational times of TAA* and RTA*

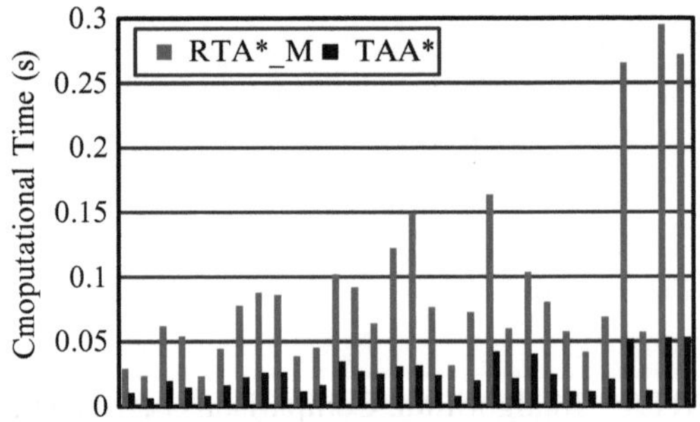

Fig. 14.8 Comparison of computational times of TAA* and RTA*_M

taken future traffic status into account in advance, but comparatively, the RTA*_M algorithm must execute the computational procedure again and again because each run takes only "short views" ahead.

14.6 Conclusions

This paper suggests a time-dependent network model, where travel speed instead of travel time is regarded as changing at different time intervals, and from which an FIFO-satisfied computational function of link travel time is derived. During the study described in this paper a temporally adaptive A* algorithm was developed for calculating least-time paths on a defined time-dependent network, where the link-travel-time computation is used as the evaluation function, and the Euclidean distance divided by the maximum possible travel speed is designed as the heuristic evaluator. An experiment on a real road network showed that the proposed algorithm is capable of foreseeing and bypassing forthcoming traffic congestion, saving travel time, and raising the overall efficiency of the navigation system.

Implementation of the TAA* algorithm needs the support of link travel speed data and turn delay data. In this study, the simple use of historical link travel speed data was used as estimates of future link travel speeds. In addition, turn delays between links were set empirically rather than theoretically. Therefore, future work will be concerned with accurate and available link travel time prediction methods and turn delay measurements and modeling in the near future.

Acknowledgements The work described in this paper was supported by the National Natural Science Foundation of China (40871184; 41001232), the China Postdoctoral Science Foundation (20090450563) and the National High Technology Research and Develop Program of China (2007AA12Z241).

References

Chabini I, Lan S (2002) Adaptations of the A* algorithm for the computation of fastest paths in deterministic discrete-time dynamic networks. IEEE Trans Intell Transport Syst 3(1):60–74

Cho HJ, Lan CL (2009) Hybrid shortest path algorithm for vehicle navigation. J Supercomput 49 (2):234–247

Fawcett J, Robinson P (2000) Adaptive routing for road traffic. IEEE Comput Graph Appl 20 (3):46–53

Fu L, Sun D, Rilett LR (2006) Heuristic shortest path algorithms for transportation applications: state of the art. Comput Operat Res 33(11):3324–3343

Gao S, Lu F (2008) An arc-labeling shortest time path algorithm. Geo-Info Sci 10(5):604–610 (In Chinese)

Hart EP, Nilsson NJ, Raphael B (1968) A formal basis for the heuristic determination of minimum cost paths. IEEE Trans Syst Sci Cybern SSC-4(2):100–107

Horn M (2000) Efficient modelling of travel in networks with time-varying link speeds. Networks 36(2):80–90

Huang B, Wu Q, Zhan FB (2007) A shortest path algorithm with novel heuristics for dynamic transportation networks. Int J Geogr Info Sci 21(6):625–644

Ishikawa H (2005) Development of a local storage type traffic prediction for car navigation system. In: The 12th World Congress on intelligent transportation system, San Francisco

Kanoulas E, Du Y, Xia T, Zhang D (2006) Finding fastest paths on a road network with speed patterns. In: Proceedings of the 22nd international conference on data engineering (ICDE), Atlanta

Kaufman DE, Smith RL (1993) Fastest paths in time-dependent networks for intelligent vehicle-highway systems application. IVHS J 1:1–11

Kim J, Jung S (2002) A dynamic window-based approximate shortest path re-computation method for digital road map databases in mobile environments. In: Shafazand MH, Tjoa AM (eds) EurAsia-ICT 2002, vol 2510, Lecture notes in computer science. Springer, Berlin/Heidelberg, pp 711–720

Klunder GA, Post HN (2006) The shortest path problem on large-scale real-road networks. Networks 48(4):182–194

Lu F, Guan Y (2004) An optimum vehicular path solution with multi-heuristics. In: Bubak M et al (eds) ICCS 2004, vol 3039, Lecture notes in computer science. Springer, Berlin/Heidelberg, pp 964–971

Lu F, Lu D, Cui W (1999) Improved Dijkstra algorithm based on quad-heap priority queue and inverse adjacent list. J Image Graph 4A(12):1044–1050 (In Chinese)

Sung K, Bell MGH, Seong M, Park S (2000) Shortest paths in a network with time-dependent flow speeds. Eur J Operat Res 121(1):32–39

Yamane K, Endo Y, Fujiwara J, Machii K (2004) Statistical traffic information for navigation system. In: The 11th World Congress on intelligent transportation system, Nagoya, Aichi

Zeng W, Church RL (2009) Finding shortest paths on real road networks: the case for A*. Int J Geogr Info Sci 23(4):531–543

Chapter 15
Evaluation of Online Itinerary Planner and Investigation of Possible Enhancement Features

Homing Tam and L.S.C. Pun-Cheng

Abstract Current information available to tourists visiting Hong Kong seems to be abundant but fragmented. Usually, individual tourists tend to plan their itinerary before they arrive, either through tourist guides or online information sources (e.g. existing itinerary planning websites). With the enhancement of information technology, it is expected that online itinerary planning will be completely supplementary to hard copy travel guides and magazines in the future. However, current online itinerary planners overlook transportation links and optimum travel planning. For example, several tourist spots can be visited sequentially and hence optimise the time available. And this is the sphere on which this project focuses. The aim is to develop a scheduling algorithm based on the Greedy Algorithm that helps prepare an itinerary for individual tourists visiting Hong Kong. Tourists' limited knowledge of the spatial extent of transportation facilities in Hong Kong has always and naturally presented obstacles as regards arranging an efficient (optimal) travel plan. With a view to minimizing travel time and maximizing the time for sightseeing, the shortest travelling time between tourist locations spots has been adopted as the guiding principle in deriving solutions. The benefits brought by the existence of this system and the currently available tourist information (both transport and tourist spot) are assessed and evaluated.

15.1 Introduction

This chapter gives the background, objectives and scope of the system. The information enables readers' understanding of the architecture of the system.

H.M. Tam (✉) • L.S.C. Pun-Cheng
Department of Land Surveying and Geo-Informatics, The Hong Kong Polytechnic University
e-mail: tamhoming@gmail.com; lspun@polyu.edu.hk

A. Yeh et al. (eds.), *Advances in Spatial Data Handling and GIS*,
Lecture Notes in Geoinformation and Cartography,
DOI 10.1007/978-3-642-25926-5_15,

15.1.1 Background

Abundant information is currently available to tourists visiting Hong Kong. However, it is rather fragmented: with information on transport and tourist attractions available on different web sites. For instance, the Hong Kong Tourism Board established a website entitled "Discover Hong Kong" some time ago. It provides information about tourist spots, accommodation to dining locations and shopping. The information available on such sites is quite diverse in content. The tourist must himself integrate the data received He may initially, have no idea how to travel from one tourist spot to another, but in general may finally work out an itinerary that does take in all his intended destinations, but yet maybe one which is not necessarily cost efficient.

Sequencing a time and cost efficient itinerary can be a difficult task, especially when the tourist is unfamiliar of the complex transportation alternatives and the spatial proximity of their desired destinations. This project aims to provide a solution enabling tourists to schedule the optimal path enabling them to visit their desired locations in an efficient sequence and to save time and money. Usually it is the limitation of time of stay in Hong Kong which makes every second become of vital importance. There is no doubt that different tourists visit the same list of tourist spots in different sequences. Optimizing the sequence of visits and preparing transport suggestions would be a valuable service.

15.1.2 Objectives

The development of a scheduling algorithm that helps prepare an itinerary for tourists visiting Hong Kong individually rather than as part of an organized tour. Tourists select the tourist spots they wish to visit. The system would propose an itinerary that includes all the user-selected spots, transport means and time taken for each leg. A solution would be derived which includes all the tourist spots selected by the user, the proposed itinerary would also include suggested transport means (and also the times taken for each journey/leg), time spent at each tourist spot and so on. In short, the objectives of the system are:

1. To provide a tool to help tourists schedule their visit
2. To provide an itinerary solution
3. To provide transport information relating the tourist spots and a particular chosen hotel
4. To provide maps of the vicinities of the points of interest (both tourist spots and hotel)
5. To suggest times spent at particular tourist spots

The ultimate objective of the system is to serve as a complete substitute for all other tourist information sources so that users can develop and print out their own itinerary.

15.1.3 Scope of the Study

Hong Kong is a city with a well-developed transport network which includes many transport modes, such as railway, bus, mini-bus, tram and taxi. Given two places, usually more than one type of transport between them is available. As the aim of the study is to investigate the tourist scheduling algorithm, the route chosen is restricted to the shortest possible travelling time.

The system must cater for a selected number of visits to such as tourist spots and hotels. It is not possible to include all of the many possible tourist spots in Hong Kong. It also should be remembered that the decision to, classify any place as a "tourist spot" is a subjective one. This pilot study system, therefore, limits the number of tourist spots selected, yet, spreads those throughout the territory. It is not simply the number of tourist spots that matters, but also the spatial extent (in terms of travelling time between the "spots"), especially when this involves cross-district travelling. For these reasons the system developed is restricted to:

1. A limited number of tourist spots.
2. Routes with the shortest travel time between tourist spots.

Other criteria (other than travelling time and cost) should also be taken into account. The rationale behind the scheduling algorithm is discussed below.

15.2 Literature Review

This section gives answers to the simply stated but difficult question: "What constitutes an itinerary planner?"

15.2.1 Itinerary

Itinerary is highly practical value to tourists. These may include geographic locations, transport, shopping, dining and any other information about the tourist spots limited to a particular area. An itinerary may also solely refer to the means of getting from one place to another. The term "itinerary" in this paper is defined as one which tourists can rely on during their stay in Hong Kong.

- *Information provided.* "Lonely Planet" is a renowned travel guidebook brand. In the book "Lonely Planet Hong Kong & Macau (City Guide)", the key contents include eating, shopping, entertainment, sports and activities and transport (Lonely Planet 2008). This classification is sensible and reasonable – the activities available at tourist spots have always been of utmost concern to tourists. The system developed should serve this principal function.

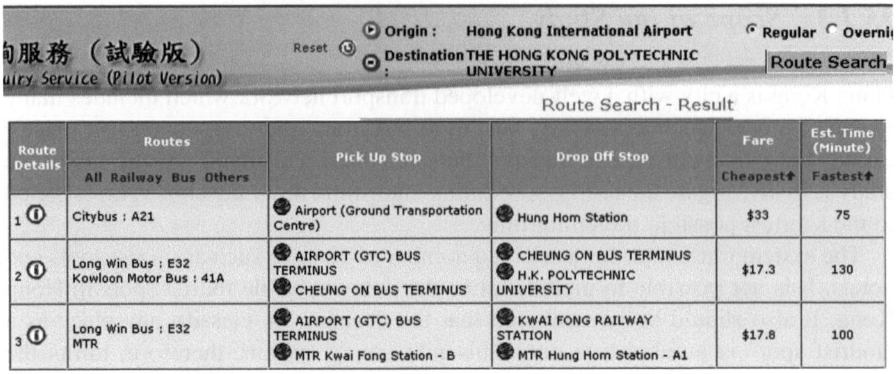

Route Details	Routes All Railway Bus Others	Pick Up Stop	Drop Off Stop	Fare Cheapest	Est. Time (Minute) Fastest
1	Citybus : A21	Airport (Ground Transportation Centre)	Hung Hom Station	$33	75
2	Long Win Bus : E32 Kowloon Motor Bus : 41A	AIRPORT (GTC) BUS TERMINUS CHEUNG ON BUS TERMINUS	CHEUNG ON BUS TERMINUS H.K. POLYTECHNIC UNIVERSITY	$17.3	130
3	Long Win Bus : E32 MTR	AIRPORT (GTC) BUS TERMINUS MTR Kwai Fong Station - B	KWAI FONG RAILWAY STATION MTR Hung Hom Station - A1	$17.8	100

Fig. 15.1 Transport route suggestions by PTES

It is obvious that not every tourist purchases a guide book such as “Lonely Planet”. Instead, free resources available on the internet or through other channels are sought. Information availability is therefore assessed.

- *Tourist spot information*. The Hong Kong Government continually promotes Hong Kong to tourists and brands Hong Kong as “Asia’s World City”. The site entitled “Hong Kong Fun in 18 Districts” set up by the Home Affairs Department (HKSAR 2008) includes a list of renowned tourist spots. The Hong Kong Tourism Board (HKTB) is an organization responsible for promoting Hong Kong to the world (HKTB 2009). Since Hong Kong is always referred as a “Shopping Paradise”, the Board has set up the web site “Where to Shop”. The site includes textual information, supplemented by photographs which provide tourists with details of individual shopping areas.
- *Transport information* is crucial to a tourist. Some cities, e.g. Sydney, have detailed arrival and departure timetables for every type of transport. This is not the case in Hong Kong – only the frequency of transport (usually given as an interval, say, 8–12 min) will be shown publicly (though detailed departures exist for managerial purposes) (Pun-Cheng et al. 2007). Most transport companies in Hong Kong have posted information online for the benefit of the public. Early in 2009, the Transport Department launched the Public Transport Enquiry Service (PTES). This service provides transport route suggestions once a user specifies an origin and a destination (Fig. 15.1).
- *Itinerary planning service provider*. There are websites that provide itinerary planning services (Shapiro 2002). TripIt is one of these. It processes such as travel confirmation emails, weather and driving directions. It may be very good for cities like San Francisco where driving between spots is the usual practice. Yet, the place where the pilot study takes place – Hong Kong, however, is a small place with seven million people. Car rental is certainly not a good option for travellers in Hong Kong. The narrow and congested road conditions in Hong Kong also discourage tourists from renting a car. On the contrary, public transport is very well developed.

HKTB launched a website which provides an interface for users to plan their itineraries. The user first selects the duration of stay, and time of day for arrival and departure. The website itineraries are impracticable for the more general purposes since it provides tours broken down into morning, afternoon and night sessions only. Nor does not cater for tourists' dining and shopping needs.

15.2.2 Scheduling Algorithm

The scheduling of an itinerary that takes the shortest time is actually a practical case of the Travelling Salesman Problem (TSP). Given a list of places, a salesman must find a path visiting all the places only once. The TSP is a well-known NP-complete problem, meaning that no algorithm exists which can solve the TSP efficiently. The exact algorithm solving the TSP computes all possible permutations and looks for the one with lowest cost. This works well for a small number of cities, but becomes computationally impracticable for cases with more than 20 places.

Heuristics and approximation algorithms are therefore devised. They give good solutions quickly (Cormen et al. 1989; Evans and Minieka 1992). With these algorithms, extremely large problems can also be solved within a reasonable amount of computing time. These solutions are probably only about 2–3% poorer than the true optimal solution. The common approach with these algorithms is to first construct a possible path. Improvement on that solution using improvement algorithms, such as iterative improvement or randomized improvement methods, is necessary.

15.3 Methodology

Any online itinerary planner should provide solutions quickly and promptly since the user is too impatient to wait. Therefore, the system devised in this paper employs an approximation algorithm followed by improvement tactics so as to compute the solution quickly. Assumptions are made for better modelling of the system.

15.3.1 Assumptions

Below is the list of assumptions considered in the initial system design stage.

1. Users value use of time in Hong Kong to be of the utmost importance. They want travelling time to be as short as possible.
2. Users only need to specify the hotel stayed at and the tourist spots they would like to visit. System will then schedule the itinerary accordingly.

3. Users always start trips from their hotel. They also end trips there. The hotel specified by the user will always be set as the start and the end of the trip.
4. Users specify the start time and end time for their visits each day.
5. In the case of the time slot (duration between start time and end time) being too short for all the spots, the system postpones the remaining spots to another day, starting at the same time.
6. Users may need to know how long it takes to get from one spot to another. A suggested staying time at a particular tourist spot should also be made available.
7. Users may want to know something about the general environment in the vicinity of the tourist spots.

15.3.2 Implementation Platform

To implement the system, a free and open source cross-platform web server package – XAMPP is employed. It is a solution stack of open source software designed for web sites on servers. It is easy to install and therefore suitable for use as the platform for this study and subsequent development. Although the time taken to run the page may differ from loading a page on a WWW server, the features of the system are to be assessed. The efficiency (or the networking issue) is left for further enhancement of the project.

15.3.3 Database

- *Design.* With respect to the assumptions listed, the database schema of the system was designed accordingly. Since the basic function of the system is to schedule the itinerary and display the public transport data between tourist spots, the database developed must have fields storing the corresponding information. Three tables are implemented in the database, namely SPOT, JOURNEY and SUBJOURNEY. The three tables are designed for different purposes: (Table 15.1 and Fig. 15.2).

15.3.4 Data Preparation

- *Data Preparation.* Travelling time is the only factor to be considered by the scheduling algorithm of the system. Real transport data were collected through the PTES (see Sect. 15.2.1). Throughout the collection of data from the PTES, the fastest route was chosen since time is of utmost concern to tourists.

Table 15.1 Purposes of the three tables in the database

Table	Purposes
SPOT	Keeps details of individual tourist spots and hotels (e.g. name, district)
	Facilitates information storage and retrieval
	Latitude and Longitude are stored to enable the use of Google Map for accurate location display
JOURNEY	Stores the transport route suggestions for all combinations of origin–destination
	Time of travel using the respective transport route suggestion
SUBJOURNEY	Stores the information for individual transport route – pickup and dropoff point also given
	Details of different legs within a journey – interchange is unavoidable in some cases

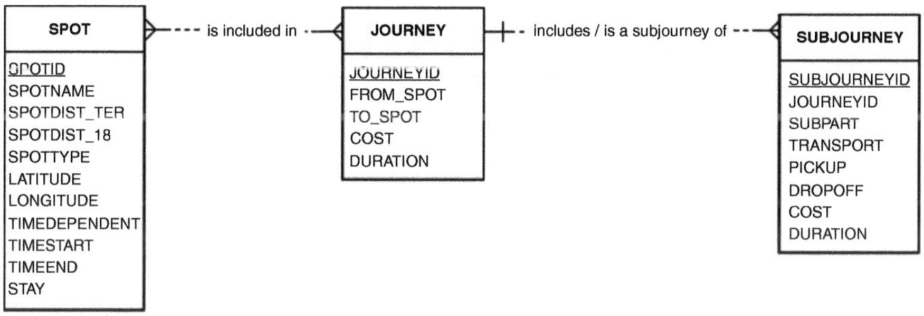

Fig. 15.2 Entity-relationship diagram of the system

Hotels and tourist spots can be treated as similar entities during the collection of transport data as they both share the properties as a node. They can be considered as Points Of Interest (POIs). Schematically, the cost (time taken) between the POIs data can be represented as a square matrix (Fig. 15.3). The first cell on the second row, labelled (2,1), holds the cost of getting from POI2 to POI1. If (2,1) is considered as the inbound journey, then (1,2) is the outbound journey for the same POI pair. If the POIs are drawn as vertices on a graph, the two edges between each POI pair represents the cost of the inbound and outbound journey between them (Fig. 15.4). The graph is asymmetric.

15.3.5 Scheduling

There are two solutions to schedule the list of tourist spots selected by the user.

1. Schedule all the tourist spots using the Greedy algorithm together with the k-opt heuristic search (General Solution)
2. Place the time constrained spots at particular positions that satisfy the constraints, then schedule the remaining spots (Time-constrained solution)

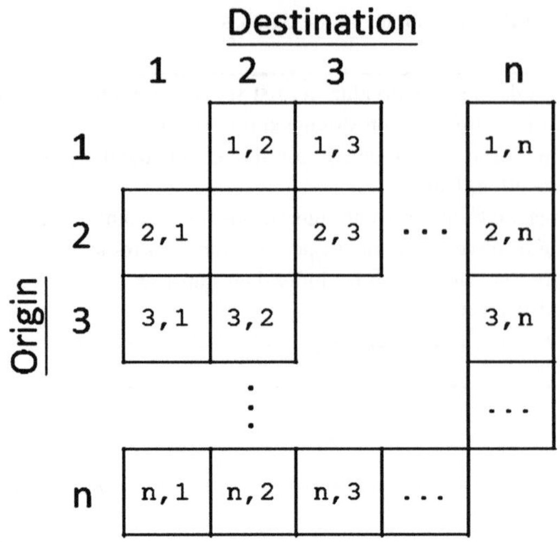

Fig. 15.3 Schematic representation of the transport routes to be prepared (*left*)

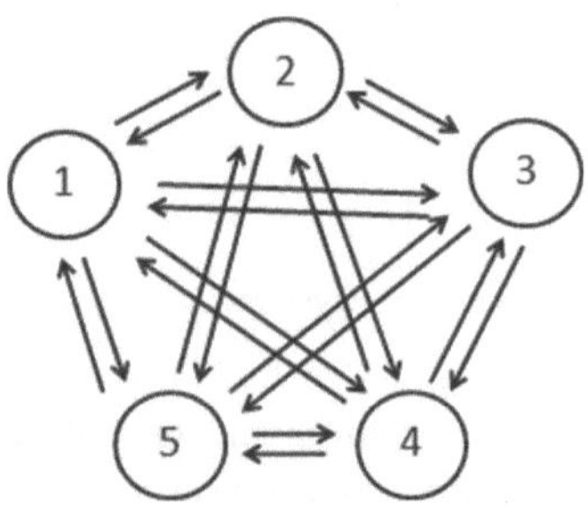

Fig. 15.4 Asymmetric graph formed upon completion of the transport route (*right*)

Table 15.2 Cost (time) matrix of possible tourist spot pairs

		Destination					
		1	2	3	4	5	6
Origin	1	/	15	10	0	10	15
	2	15	/	5	15	20	5
	3	10	25	/	10	25	20
	4	0	15	10	/	15	15
	5	15	20	30	15	/	20
	6	15	5	10	15	20	/

Solution (1) has been devised in the prototype platform while (2) is still theoretical. Real data was used for (1) and an extract of the cost matrix is given in Table 15.2 as an illustration.

Fig. 15.5 The scheduling solution with Greedy Algorithm

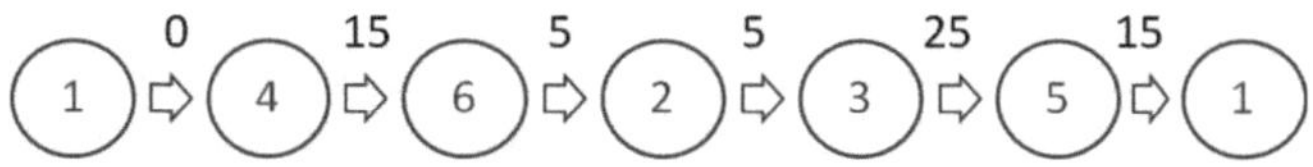

Fig. 15.6 The scheduling solution with Greedy Algorithm with tour improvement heuristic

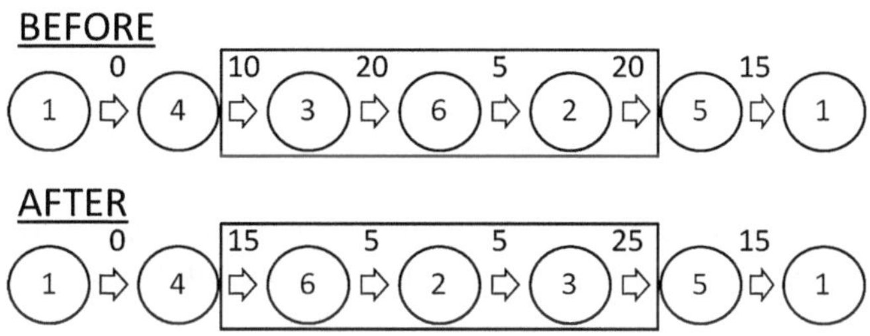

Fig. 15.7 Difference in position before and after tour improvement heuristic (3-Opt adopted)

- *General Solution.* Consider a case where the user has selected the hotel (e.g. SPOTID = 1), and some tourist spots with SPOTID (e.g. SPOTID 2-6). With the use of the greedy algorithm (GA) alone, the solution will be like Fig. 15.5. The user starts and end at the same spot (SPOTID = 7). Sequence (numbers inside circles represents tourist spots) while the figures between them represent travel time.

Since there is a possibility that the path generated, may be the longest path (Bang-Jensen et al. 2004), the solution needs to pass through an improvement step. The tour improvement heuristic is adopted as the improvement tool. The result generated with that improvement tool applied is (Fig. 15.6).

Attention is to be drawn to the difference in position of the three spots located in the middle before and after the application of the tour improvement heuristic. The difference in position of the three spots is marked by the rectangles in Fig. 15.7.

The resulting difference in Fig. 15.7 is actually the result of the K-Option improvement heuristic. For this case, the value of K was set as 3. The repositioning took place since a shorter path is available (note only the figures above the arrows in both rectangles). There should be at least five spots in the itinerary (four if the start and end point are the same). The first spot and the last spot are fixed (because of their nature). An exhaustive list of combination starting A and ending E will be drawn, while would yield *3!* (a total of six) permutations (only three spots in between can be swapped).

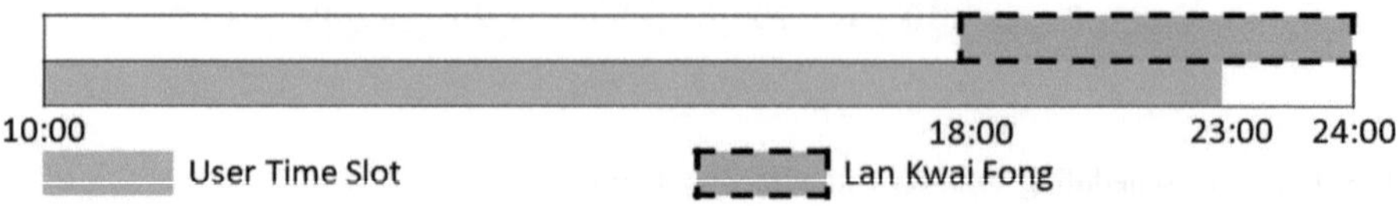

Fig. 15.8 User defined time slot and operating hours of time constrained spot

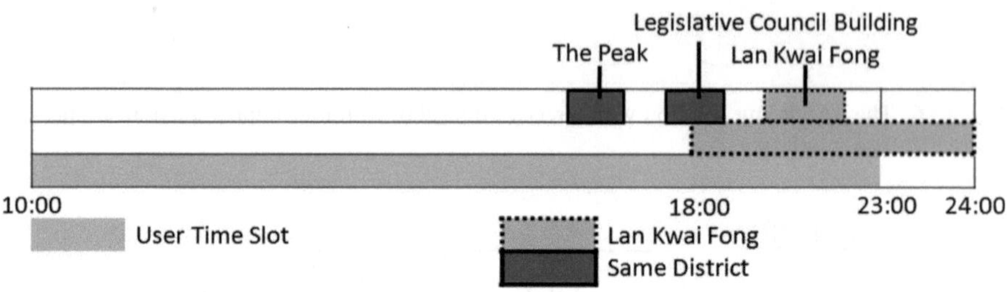

Fig. 15.9 Time constrained spot with some same district spots inserted

For a list with more than five spots, the 3-Opt will be first performed on the second, third and four spot. Then the floating window will be shifted to the third, fourth and fifth positions. The algorithm proceeds with its processing until the last of the floating window touches the *(n−1)th* spot.

By illustrating how the floating window works, the difference in sequence generated solely by the greedy algorithm and greedy algorithm with 3-Opt employed is now justified.

The reason for setting the K value as 3 in the improvement heuristic is due to the spread of the tourist spots over different districts – there are about three tourist spots in the same district in the database. Since it is rather unlikely that the travelling time will be improved (in terms of reduction in total travelling time) by including distant tourist spots (i.e. tourist spots that located in different districts), the value 3 is justified.

- *Time-constrained Solution*. The time slot fitting approach algorithm deals first with time constrained tourist spots. To make it easier to understand, the illustrations include ONE time constrained spots only. Lan Kwai Fong is selected as the time constrained spot to demonstrate this approach. It is assumed that the bars in Lan Kwai Fong only operate from 18:00 to 24:00 daily (Fig. 15.8).

Lan Kwai Fong would then, be inserted as late as possible in the user defined time slot (Fig. 15.9). Visit to the Lan Kwai Fong should finish some time before the user defined time slot since it takes some time for the tourist to return to the hotel. Tourist spots in the same district will then be inserted before the time slot assigned to Lan Kwai Fong (Fig. 15.9).

Once the tourist spots (in the same district) have been inserted, the algorithm looks for the possibility of making the whole trunk being switched to an earlier time (Fig. 15.10). The rationale for this move is because there is a chance that the schedule is subject to delay (by traffic congestion). If there is a delay of 30 min

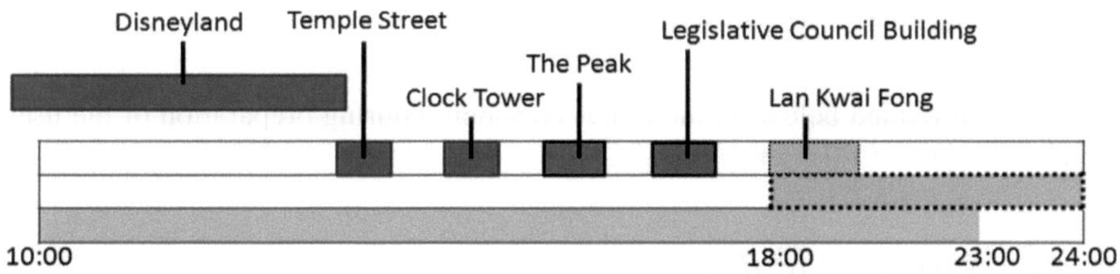

Fig. 15.10 Time constrained spot and two other spots shifted forward

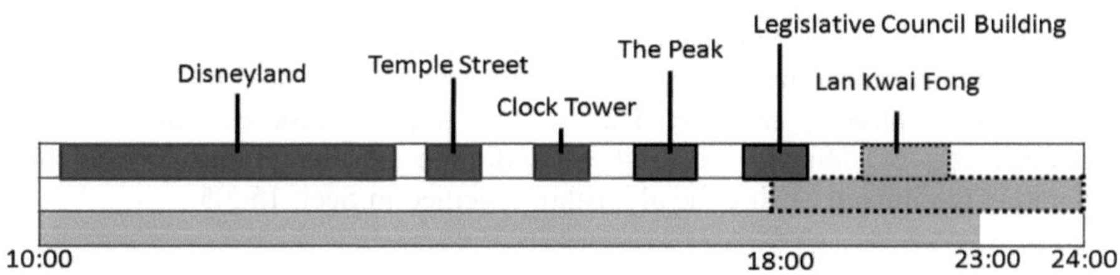

Fig. 15.11 Forward shift of the whole route in order to fit in an extra tourist spot

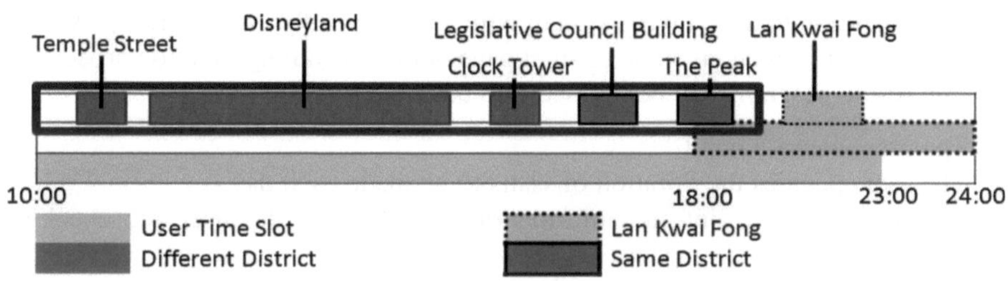

Fig. 15.12 Optimization completed for the spots before the time constrained spot

for the day, the maximum time that the user can spend in Lan Kwai Fong would reduce by 30 min.

Once the shifting is completed, tourist spots for other districts will be inserted before the scheduled tourist spots. If the amount of time needed to finish the trip is longer than the current time available. The system looks for the possibility of a shift to a later time (Notice that the time visiting the time constrained spot has been shift earlier, so, there is some room for another shift).

Extra tourist spots are then fitted in (Fig. 15.11). Optimization takes place by implementing the k-opt heuristic iterative search. The sequence prior to the time constrained spot is said to be fixed after the optimization (Fig. 15.12). It should be noted that the swapping in position between Temple Street, Disneyland and Clock Tower is fictional. This only serves to illustrate the optimization (3-Opt) only.

The scheduling of spots later than Lan Kwai Fong will then be continued. The scheduling mechanism is much simpler. The time constrained spot (Lan Kwai Fong) is the start point and the hotel is the end point.

15.4 System Implementation

Details are provided below on the issues considered during preparation of the user interface and other relevant aspects.

15.4.1 System Workflow

Four steps are taken in building the customized itinerary (Fig. 15.13). The Hotel should be chosen first since it is the first and last spot of the itinerary. Secondly, the arrival and departure time lets the system know how much time the user will be available for sightseeing each day. The user then selects the tourist spots. Once the user clicks the "Schedule" button and they will processes the information and then generating the itinerary using the algorithm specified in Sect. 15.3.5.

15.4.2 User Interface and Report Printing

It is the intention to develop the system with a user-friendly user interface. A short text description will be given on the first page of the scheduling function procedure.

The whole program will be divided into three steps:

1. The user selects a hotel.
2. The user specifies the duration of visits to be made each day.
3. The system will print the information specified by the user for checking (Fig. 15.14).
4. The user selects a number of tourist spots.

The user may then select the spots and leave the system to do the scheduling. Figures 15.15 and 15.16 show a typical layout of the itinerary generated. The system is also capable of handling exceptional cases. For example, an ambitious user may select numerous tourist spots that cannot be visited within the time slot specified. The system alerts the user to this fact while scheduling the trip for another day (Fig. 15.17).

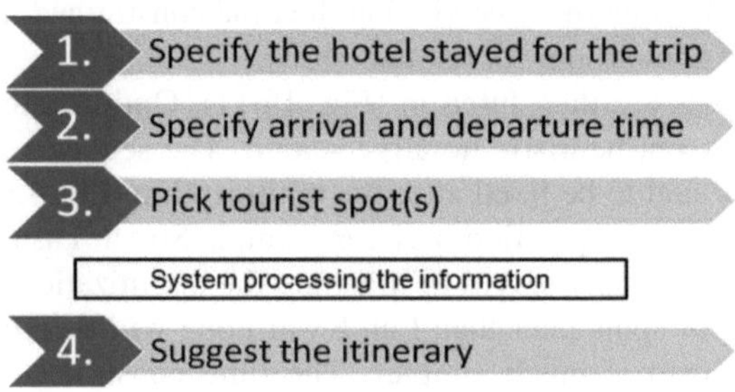

Fig. 15.13 The four-step system workflow of the database

HOTEL CHOSEN		START TIME AND END TIME	
Hotel Name:	The Pennisula	Start Time:	10:00
Territory:	Kowloon	End Time:	20:00
District:	Yau Tsim Mong	Duration:	10 hour(s) 00 minute(s)

Fig. 15.14 Information shown upon the completion of step 1 and 2

SCHEDULING RESULT

HOTEL INFORMATION	
Hotel Name :	The Pennisula
Territory :	Kowloon
District :	Yau Tsim Mong

YOUR TRAVEL PLAN	
Start Time:	10:00
End Time:	18:00
Duration:	08 hour(s) 00 minute(s)

SUMMARIZED SCHEDULE (IN SEQUENCE)

1. Temple Street
2. Golden Computer Arcade
3. Wong Tai Sin Temple
4. Legislative Council Building
5. The Peak

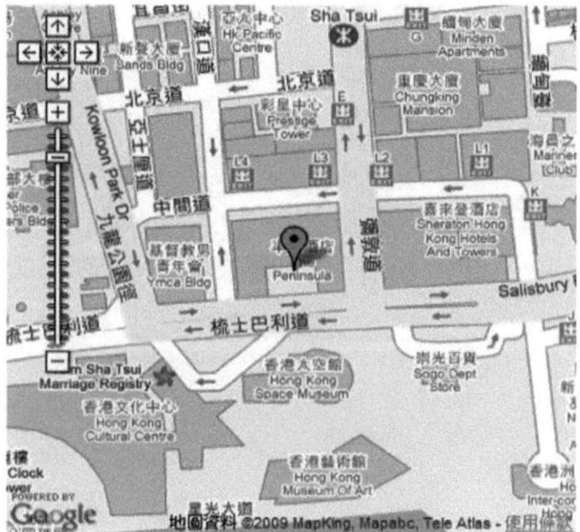

Map of the hotel (Click to enlarge)

Fig. 15.15 Itinerary generated (extract)

15.5 System Evaluation and Enhancement

15.5.1 System Performance

The system receives user input and generates the path without any error. Yet, the execution time is not recorded. The small size (in terms of number of tourist spots) does not put much stress on the time of computation time.

Since there is only one user in the platform and concurrent user under development platform, it is rather hard to perform a stress test or any similar type of testing. Normally, the system is able to return a result within 5 s for a case where all the tourist spots have been selected. Around 500 database transactions had taken place.

15.5.2 System Limitation and Possible Enhancement

- *Programming language*. PHP is easy to program and implement. It is possible for later system to consider the compilation time for different languages and choose a better and more efficient one. Compatibility between the scripting language and the server support should also be considered.

11:10 - 11:30 Arrive at **Golden Bauhinia Square (Wan Chai)** and do the sightseeing.

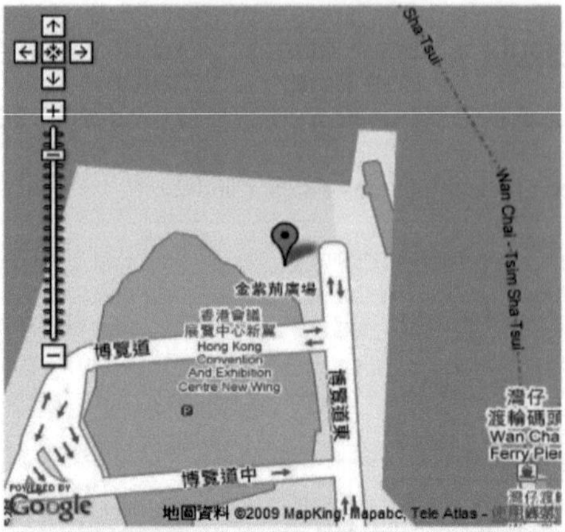

11:30 - 12:10

STEP	TRANSPORT	PICKUP	DROPOFF
1	Kowloon Motor Bus / Citybus : 117	CONVENTION AVENUE	KWEILIN STREET SHAM SHUI PO

12:10 - 13:10 Arrive at **Golden Computer Arcade** and do the sightseeing.

Fig. 15.16 Detailed transportation means tourist spots

18:10 - 19:50

STEP	TRANSPORT	PICKUP	DROPOFF
1	Kowloon Motor Bus : 64K	FONG MA PO (LAM TSUEN WISHING TREE)	TAI WO RAILWAY STATION
2	MTR	MTR Tai Wo Station - Tai Wo Indoor Recreation Centre Access	MTR East Tsim Sha Tsui Station - L3

ATTENTION:
You have selected too much tourist spot to fit in your desired time slot of 1 day(s).
You will continue to carry out on Day 2 again.

09:00 - 09:55

STEP	TRANSPORT	PICKUP	DROPOFF
1	MTR	MTR Tsim Sha Tsui Station - F	MTR Sai Wan Ho Station - A
2	New World First Bus : 14	Sai Wan Ho Station, Tai On Street	Stanford Villa, Stanley Village Road

Fig. 15.17 Prompt of overtime visiting

- *Development Platform and Implementation Platform.* In this project, a combination of PHP and MySQL was used to implement the scheduling. It should be noted that this combination was adopted because of the convenient and readily available technical support.
- *Scheduling algorithm.* The rationale behind setting the K value of the K-opt heuristic as 3 is discussed above. This "magic number" is to be revised once necessary. It should be noted that careful selection of the k value is needed and should be limited to a small number as the number of combinations increases geometrically. The system currently uses the greedy algorithm together with a heuristic search to seek for possible improvement of the result. Though the problem size in this study was small enough for an exhaustive search, it is not the intention of the system to compute the shortest solution, but rather an optimal solution. The real time computation of the total travelling time for an exhaustive search possesses would take longer for the system to complete (the schedule), because of too many queries to the database.

To avoid making an exhaustive search in deriving an optimal solution, an alternative is to screen out "impractical cases" from the exhaustive search. For example, given five tourist spots, the tourist spots can be further grouped by district so as to filter out impracticable combinations.

- *Solution with time constrained spots.* There is a time limitation and difficulty in terms of coding in the implementation of the time constrained solution. The current scheduling ignores the tourist spots set times: e.g. opening hours (e.g. the Hong Kong Disneyland), operating hours (e.g. the Museums), or preferred hours (e.g. visiting the Peak at night is preferable to the afternoon). The algorithm to solve such as time constrained spots is discussed above.
- *Which weighs more: time or distance?* The pilot study only considers predetermined travelling times when creating a schedule. It is, however, a fact that traffic conditions differ greatly between peak and non-peak hours. A distance factor may be applied to check the chances of traffic congestion during peak hours.
- *From the nearest district to the farthest district?* The system currently prepares the initial solution using the Greedy Algorithm, which causes the resulting itinerary to visit the nearest spot first. That means the solution will begin with short transport times and progress to longer transport times. It is doubtful, however, that the total time of the trip would be less if the trip started the other way round – i.e. the system first picks the furthest spot from the origin, performs the greedy algorithm and then the improvement heuristics.
- *Mapping of results.* Currently, the map only shows the location of each spot. The map does not provide detailed information to enable the user to get a comprehensive idea of the transport facilities that the system suggests.

15.5.3 Suggestions

The computation of the initial solution (which needs to be improved upon in the future) is derived from the greedy algorithm. Some studies have pointed out the greedy algorithm may not come up with a good solution, or even a reasonable solution for later improvement. The choice of the algorithm was mainly due to its simplicity. Further enhancement should explore the possibility of using geographical relationships as an aid to scheduling and better utilization of online map resources. This may help the user learn more about the surrounding environment.

15.6 Conclusions

As stated in the beginning, the purpose of the project is to devise a system capable of scheduling an itinerary that satisfies the conditions specified by the users. The scheduling algorithm and the database architecture are comparatively simple. It is expected that further enhancement can be made to both the algorithm and the database architecture. The use of hardware to accommodate/suit the needs of the system is to be assessed and further developed in the future.

Appendix

Implementation of the pilot study is available at: http://www.iplan.hk.

References

Bang-Jensen J, Gutin GL, Yeo A (2004) When the greedy algorithm fails. Discrete Optimization 1:121–127

Cormen TH, Leiserson CE, Rivest LR (1989) Introduction to algorithms. MIT Press, Cambridge, MA (ISBN-10: 0262031418)

Evans RJ, Minieka E (1992) Optimization algorithms for networks and graphs, 2nd edn. CRC Press, New York (ISBN-10: 0824786025)

Home Affairs Department, HKSAR (2008) Hong Kong fun in 18 districts. Hong Kong Special Administrative Region. http://www.gohk.gov.hk/. Accessed 31 Mar 2009

Lonely Planet Hong Kong, Macau (City Guide), (ISBN-10: 1741792258)

Public Transport Enquiry Service (Pilot Version) http://ptes.td.gov.hk/. Accessed 27 Nov 2009

Pun-Cheng L, Mok ECM et al. (2007) EASYGO - a public transport query and guiding LBS. Lecture notes in geoinformation and cartography, Location based services and teleCartography. Springer, pp 545–554, ISBN-10: 3540367276

Shapiro M (2002) Internet travel planner: how to plan trips and save money online. Globe Pequot Press, Guilford. ISBN (ISBN-10: 0762712171)

The Hong Kong Tourism Board (2009) Discover Hong Kong. http://www.discoverhongkong.com. Accessed 1 Feb 2009

TripIt – Online travel itinerary and trip planner (2006) http://www.tripit.com/. Accessed 27 Nov 2009

Part VI
Application of GIS and Remote Sensing

Chapter 16
A Decision Support Framework for the Risk Assessment of Coastal Erosion in the Yangtze Delta

Xing Li, Yunxuan Zhou, Fang Shen, Runyuan Kuang, Wen Wu, and Zongsheng Zheng

Abstract Coastal erosion is an issue of widespread concern and the Yangtze Delta is one of the most important economic regions in China, has been showing a trend towards such erosion. Possible influences are global warming in the form of sea level rise and the increasing human activities in the catchment and its estuary. Most Current studies on coastal erosion in this region have focused on the causes and types of erosion. This paper presents a Decision Support Framework (DSF) to estimate the likelihood and possible degree of coastal erosion in the coastal zone of the Yangtze Delta. The framework also enables consideration of the potential management problems and economic development challenges. It has four major components: an integrated database, GIS-based risk assessment models, scenario generator and a visualization toolkit. Specific focus is on the development of a GIS-based risk assessment model for the muddy coasts of the Yangtze Delta. An Analytic Hierarchy Process (AHP) method was used to weight the variables of the model, such as coastal elevation, coastal slope, et al. The assessment results show the validity of the approach. The DSF will make relevant specialized data and information more accessible to managers, and has an extensive capability to facilitate communication and subsequent synergetic work between man and computers.

X. Li (✉)
School of Geodesy and Geomatics, Xuzhou Normal University, Xuzhou 221116, China
e-mail: li_xing99@yahoo.com.cn

Y.X. Zhou • F. Shen • W. Wu
State Key Laboratory of Estuarine and Coastal Research, East China Normal University, Shanghai, China

R.Y. Kuang
School of Architectural and Surveying & Mapping Engineering, Jiangxi University of Science and Technology, Ganzhou, China

Z.S. Zheng
College of Information, Shanghai Ocean University, Shanghai, China

A. Yeh et al. (eds.), *Advances in Spatial Data Handling and GIS*, Lecture Notes in Geoinformation and Cartography,
DOI 10.1007/978-3-642-25926-5_16,

16.1 Introduction

Coastal erosion is an issue of widespread concern. It is estimated that at least 70% of sandy beaches in the world is retreating at a rate of 0.5–1.0 m/year (Bird 1985). It is argued that intensified human activities and accelerated sea level rise will aggravate the coastal erosion in the future century (Zhang et al. 2004). The coasts, especially of the deltas and megadeltas, already recognized as being highly susceptible to human and natural impacts, will be increasingly exposed to risk (Nicholls et al. 2007). In most Chinese coastal areas, erosion is ubiquitous (Sheng and Zhu 2002; Xia et al. 1993). Zuo et al. (2009) reported that most of the coasts between the mouth of the Luan River and the city of Qinhuangdao (northeast Bohai Sea, China) have suffered from extensive erosion with an average rate of 3.7 m/year during 1986–2000. The abandoned Huanghe delta has been subjected to dramatic erosion since 1855 when the channel of the Huanghe River shifted (Zhang et al. 1998). As far as the Yangtze Delta is concerned, the huge sediment load from the Yangtze River has contributed to successive accretion. Contemporary global warming effects and the construction of large hydraulic engineering works in the catchment and its estuary significantly decelerated the trend, and the delta degradation has been observed since 2003 (Yang et al. 2007). The combined influence of planned dams and the South-to-North Water Diversion Project will make the anticipated coastal erosion problem even worse.

Since the Yangtze Delta is one of the most important economic regions in China, long-term coastal spatial planning is an inevitable issue for local governments. Most coastal areas in the Yangtze Delta are typically muddy and are characterized by extremely complex land-sea interaction. Human and natural dimensions under the circumstances of global change are complex and uncertain, hence there is a need for such as a Decision Support Framework (DSF) to integrate human and computer intelligence. Although decision support has been involved in integrated coastal Zone management (ICZM) early in the 1990s, it has a limited application in today's conditions (Van Kouwen et al. 2008). Most of the existing decision support related to coastal erosion is potentially underlain by the complete physical, socio-economic datasets and the deterministic solutions. Of interest is that in many developing countries some key parameters affecting model outputs are unavailable (Szlafsztein and Sterr 2007), causing relevant research, in these regions, is strictly limited. In the case of the Yangtze Delta, most current studies about coastal erosion still focus on the causes and types of erosion (Cai et al. 2009; Chen and Zong 1998; Gao and Wang 2008; Hori et al. 2002; Milliman et al. 1985; Yang et al. 2001; Yang et al. 2006). As indicated above, this paper presents a DSF for the risk assessment of coastal erosion, in an effort to be of help in the consideration of the potential management problems and challenges for economic development in the coastal zone of the Yangtze Delta.

The primary goal of this study is to develop a framework to provide a theoretical and methodological base for future practical applications, and to further bridge the gap between scientists and managers to provide an intelligence base for scientific coastal management.

16.2 Why Not Decision Support System (DSS)?

The idea of DSS originated in the mid-1960s (Power 2008), and has been applied extensively in many fields. In comparison with work in other areas, its application in coastal management is limited (Wiggins 2004). The causes of this limitation have been discussed in the Literature (cf. Jones et al. 2002; Uran and Janssen 2003; Van Kouwen et al. 2008; Westmacott 2001). The specific term DSS in this study is avoided for the following reasons.

Nowadays, the development of computer technology has been responsible for few computer systems have proved to have insufficient functions to adequately support necessary decision making (Alter 2004). Hence it is felt by the authors that DSS may be a still popular but has little meaning for specific application areas.

The technological and administrative background of the Yangtze Delta has not yet used DSS for coastal management (Lau 2005; Shi et al. 2001). China is a developing country, and the Yangtze Delta is one of its economic centers. The local authorities have put most of their energy and focus into economic development scenarios. Attention to scientific coastal management appears to have been neglected. The potential economic pay back has not been understood. There is little doubt, however, that as regards the long term effects, scientific coastal management has prime importance. Neglect of this and the corresponding effects of climate warming have the potential for a negative economic impact on the whole area. Hence specific technology is needed both to identify the key issues confronting coastal managers and also to contribute to their eradication.

Coastal erosion is a complex natural phenomenon, especially as regards muddy coastlines. Although information technology is quite advanced, decisions still mainly depend on experienced specialists and managers. In the first instance, the computer system used in this area needs to be improved by incorporating more sophisticated models. However, before any changes can be made, it must be remembered that the understanding of nature is a gradual process, in part, because of the very many unrecorded unknowns, an example being the complexity of coastal erosion. Although a great number of research studies have been conducted on the Yangtze Delta over the past decades, knowledge, appropriate to today's conditions, is very limited. Clearly, a tool or platform is needed to further facilitate the understanding the natural and human-induced coastal behavior. As indicated, DSS, as a technical artifact, is not the answer, as it cannot enable an explicit expression of the above intricacies problem scenario.

Because of the above considerations, a "decision support framework" is used. It has the following three characteristics: (1) ability to provide reference data to promote the effectiveness of decision making; (2) the ability to provide analysis models to promote the efficiency of decision making; (3) the ability to bridge the gap between scientists and policy-makers by providing the specific scenario and related visualization tools. The end product is the facilitation of the "coevolution" of human and computer as regards scientific coastal management.

16.3 Decision Support Framework Challenges in the Yangtze Delta

The Yangtze Delta is formed mainly by riverine sediments from the Yangtze River and arises from the river discharge and ocean processes (Chen et al. 1959) (Fig. 16.1). The coasts are mostly muddy, and the muddy tidal flats are extensively distributed and have varying widths. The coastal development and make-up is complex, being influenced by many factors, such as its specific coastal evolution pathway, ocean dynamic, terrigenous materials, human activities, socio-economic conditions.

The Yangtze Delta is a tide-dominated coastal environment (Saito et al. 2001). The mean annual water discharge of the Yangtze River is 9.24×10^{11} m^3 that transports 4.80×10^8 t of sediment to the sea (Yang et al. 2006). The water discharge has an evident yearly and seasonal variation (Chen et al. 2001). The annual tidal prism into the Yangtze estuary has a total volume of 8.40×10^{12} m^3, which is an order of magnitude higher than the water discharge (Chen et al. 1988). Along the coasts, the semidiurnal tide is predominant with different tidal ranges. The average tidal range is 1.5–1.7 m around the abandoned Yellow River mouth and along the southern part of the North Jiangsu coast (Zhou et al. 1994). The mean

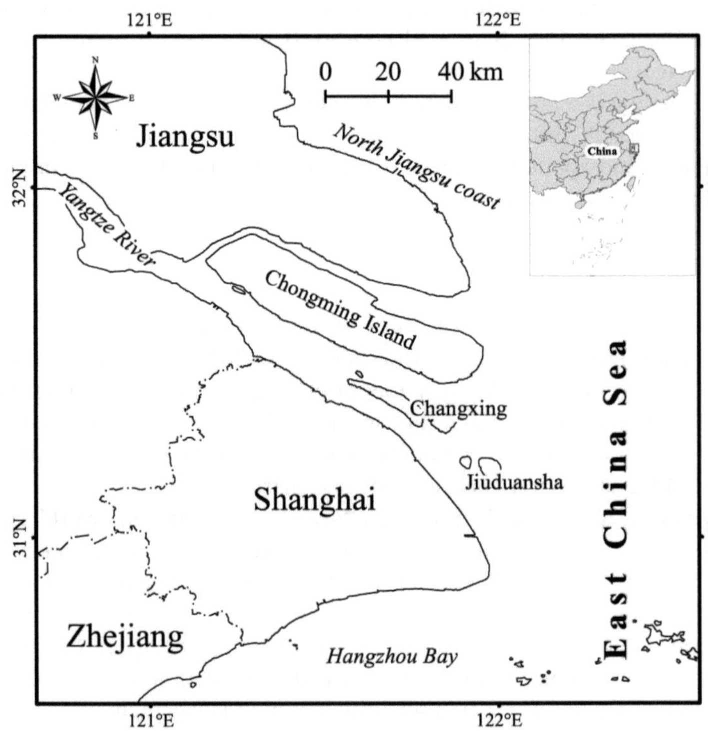

Fig. 16.1 Location of the Yangtze Delta

tidal range is 2.7 m near the Yangtze River mouth, and the maximum tidal range approaches 4.6 m (Shen et al. 1988).

The Yangtze Delta is one of the most important industrial and agricultural areas in China where economic development is the first priority of local governments. With 2% of the area and about 10% of the population of China, the region contributed to 26% of national GDP in the first half of 2008 (http://www.ocn.com.cn/reports/2006079changsanjiao.htm, in Chinese). Accompanying the urbanization and industrialization development will be the reclamation of more tidal flats of the Yangtze Delta. The result will be a growing disturbance and subsequent vulnerability of the coastal environments. The delta is one of the most serious subsidence areas in China (Han et al. 2008). During 2001–2006, the mean subsidence rate was 12.7 mm/year in Shanghai (Gong 2008). The fluvial sediment to the coastal ocean will continually reduce as large hydraulic engineering works are settled within its catchment. Additionally, the sea level is rising at a significant rate under the influence of global warming (Shi et al. 2001). It is estimated that in the Shanghai Municipality, the relative sea level rise from 1990 to 2050 will be 50 cm, 20–30 cm on the coastal plain of North Jiangsu and the northern bank of Hangzhou Bay, and the 100-year storm surge level will increase by 38–44 cm when the sea level reaches 50 cm (Shi et al. 2000). These statistics illustrate the almost undoubted increase of coastal erosion risk in the Yangtze Delta.

16.4 The Decision Support Framework

The DSF consists of four major components: (1) integrated database, (2) GIS-based risk assessment models, (3) scenarios generator and (4) visualization toolkit. Firstly it is necessary to define the three concepts used in order to facilitate further DSF descriptions and discussions.

(a) *Visualization*: a set of computer-generated static or animated, 2D or 3D characters, graphics and images to depict the properties and process of physical objects in certain aspects, with the intention of visually communicating information to specified audiences.
(b) *Scenario*: an expression of a subset of the complete parameters that can completely depict a complex system. Scenarios are neither predictions nor forecasts, but rather an attempt to provide a picture situation under specified conditions, to enable an intuitive grasp of the essence of events or phenomena.
(c) *Decision support framework*: a conceptual structure comprising personnel and computer programs, in which manager/decision-maker and computer programs interact in decision making, in other words: "co-evolve".

The three concepts are defined based on the consideration of a highly complex, dynamic environment in which scientists have difficulty in presenting information related to scientific findings, to managers/decision-makers for their complete and effective understanding of what is needed to maintain both profitability and the corresponding protection of the environment (Tribbia and Moser 2008).

Visualization and scenarios act as tools to help those without technical expertise understand the complex physical objects and subsequently update their own knowledge. The framework contributes to combine the strengths of both human intelligence and computer programs, and promotes the evolution of both. The development of such synergy will enable the framework to improve with time.

The integrated database provides a means of storage, management and retrieval of all relevant data, which support the risk assessment of coastal erosion in the Yangtze Delta. It contains direct indicator data (can be directly input to the model as a parameter), input data (for calculations of direct indicators) and auxiliary data (to aid visualization and scenario). These data are represented respectively in raster and vector format. The involved data types have, administrative boundaries, remote sensing images, geomorphology and geology, elevation, bathymetry, shorelines, sediment transport, riverine discharge, tidal regimes, wave climates, sea level rises, subsidence, extreme historical events, Land Cover/Land Use (LCLU), infrastructure, ecological hotspots, population density.

GIS-based Risk Assessment Model is the core component of the DSF. It is used mainly to quantify the risk level of coastal erosion for the muddy coasts of the Yangtze Delta, and where possible, further serve as a general framework for other muddy coasts. The following steps are followed: (1) definition of homogeneous units; (2) the identification of the vulnerability indicators and the evaluation of vulnerability; (3) the identification of the impact indicators and hazard evaluation; (4) the risk evaluation. All the calculations are conducted in the GIS environment where each are allowed to take advantage of its individual enhanced spatial analysis capabilities.

The scenario generator is used to generate the impact of coastal erosion under the varying conditions which may exist, as hypothetic, simulated or predicted. Chiefly, it assists users to understand the coastal environment, define the vulnerability and impact indicators, and determine weights. It can especially, be instrumental in the incorporation of human knowledge and insights into decision making processes. Before a given scenario is generated, several other steps are needed. Firstly, it is necessary to plan the scenarios, such as worst case, best case or current case, in terms of the features in mind. The required data or parameters are input into relevant models to generate scenarios. Finally, different scenarios are tested, and from these scenarios, knowledge is extracted. Again, some prediction models about shoreline change and sea level rise are involved in the procedure. The visualization toolkit in the next subsection supports the entire procedure.

Visualization toolkit provides some basic visual forms. Some aspects of functions in the toolkit are especially worthwhile. The firstly is organically integrated specialized data and information on coastal erosion input to GIS. Visualization techniques and specialized computational models are developed separately. Recent research indicates that GIS has been one of the most recognizable management tools (Tribbia and Moser 2008). Hence, integrated representation is feasible and necessary. Integration provides insight into hidden patterns and trends for end users. The second is visualization of the evolution process and trends of coastal elements by predictive simulation models. The coastal elements data, such as shoreline, tidal vegetation, reclamation, nearshore sediment movement, are captured at discrete time intervals without intermediate details. To some extent, the function allows an understanding of the complete evolution process. The third is linking coastal landscapes and dynamic elements with Digital Elevation Model (DEM) by extensive texture and 3D models. Some improved interpolation algorithms, such as fractal Brownian motion (fBm), wavelet, are used to enhance the accuracy and visual effects of DEM. In addition, uncertainty of data and results, for instance, historical shorelines and predicted shorelines are presented.

16.5 Risk Assessment of Coastal Erosion in the Yangtze Delta

In this section, the details of risk assessment as regards the study area, based on the DSF, especially GIS-based risk assessment model, are given.

Risk assessment has varied definitions and the term is widely used in various fields (Del Río and Gracia 2009; Eurosion 2004; UN/ISDR 2004). Usually, a whole risk assessment encompasses vulnerability assessment and hazard assessment (UN/ISDR 2004). The former is often linked to physical dimensions to determine the area size, probability and/or intensity of occurrence of a hazard under specified conditions. The latter determines the damage potential of a hazard, and is often related to the socio-economic and human dimensions. Each has a respective dependency on a set of vulnerability indicators and impact indicators. The resulting risk index (RI), can be derived by combining the two aspects, as a single risk measurement.

In accordance with the steps described above, some primary indicators, which can represent the essential physical and socio-economic characteristics of the specified coastal system, need first to be selected to segment the shorelines into the relatively homogeneous units. These primary indicators include *administrative boundary*, *population density*, *geological types*, *and coastal characteristics*. The full segmentation procedure requires an intersection operation of four different segmentation results related to these four indicators, and finally produces 40 segments for all shorelines of the study area.

The selections of vulnerability and impact indicators are followed by some premises, including independence, availability and ease of use. According to specialists' suggestions and ideas from published literature (e.g., Feng and Xia 2003; Li and Yang 2001; Wang et al. 1999), the ten vulnerability indicators are identified as:

- Coastal elevation
- Coastal slope
- Average annual deposit volume
- Shoreline change rates
- Tidal range
- Significant wave height

- Relative sea level rise
- Intertidal width
- Intertidal vegetation type
- Intertidal vegetation zone width

and the three impact indicators are identified as:

- Population density
- Land use type
- Ecological hotspots

Both vulnerability and impact indicators are then quantified or classified as numerable variables. Based on field surveys, the most recent seawall was interpreted from the Landsat TM images of April 25, 2008, and to produce an assessment baseline (polyline feature class) in ArcGIS. Most variables were calculated in a buffer zone of a certain width, and these values corresponded to each homogeneous unit. The variable of "coastal elevation" represents the percentage of the total area with an elevation of less than 2.4 m above the mean sea level in a 5 km wide buffer zone landwards from the assessment baseline. The "coastal slope" is calculated in a 2 km wide buffer zone of 1 km landwards and seawards, respectively, from the assessment baseline. The buffer zone widths for "average annual deposit volume", "significant wave height" and "relative sea level rise" are, respectively, 2 km, 5 km and 5 km seawards from the assessment baseline. "Intertidal width" refers to average intertidal width from assessment baseline to 2 m isobaths. The main land use type, as the variable "land use type", was identified from the Landsat TM images in a 2 km buffer zone landwards from the assessment baseline. Vegetation data, interpreted from TM images based on field surveys, represent the vegetation outside the seawalls.

Subaqueous topographic data and elevation data were obtained from digitized nautical charts and 1:50,000 scale topographic map, respectively. Landsat TM images from 1990 to 2008 were used to extract shorelines. Global mean sea level data with 1/3 of a degree resolution provided by AVISO are also available at http://www.aviso.oceanobs.com/en/news/ocean-indicators/mean-sea-level/. Significant wave height was simulated by the SWAN wave model based on an 11-year (1995–2005) monthly climatological data set from NOAA/NESDIS/National Climatic Data Center website (http://www.ncdc.noaa.gov/oa/rsad/seawinds.html). Population density data were obtained from the Gridded Population of the World version 3 (GPWv3) data set with 2.5 arc-minutes resolution (CIESIN/FAO/CIAT 2005).

Depending on the nature of each of these variables, each was then assigned ranks ranging from one to five, with one representing minimum vulnerability/hazard and five representing maximum vulnerability/hazard.

Before evaluating vulnerability and hazard, the above variables must be weighted, based on their relative importance in determining coastal erosion vulnerability and hazard. The aims have two aspects: (1) to avoid underestimation or overestimation of the contribution of any variable; and (2) to incorporate specialist knowledge into the DSF. To assign the objective and reliable weights to the variables, an Analytic Hierarchy Process (AHP), has been extensively used in almost all applications related to Multiple Criteria Decision Making (MCDM) over the last 20 years (Ho 2008), was employed for the variable weightings. According to the fundamental scale of absolute numbers (Saaty 2008), the priority scales for each variable was derived based on the judgment of experts, and to construct a pair-wise comparison matrix. For the ten vulnerability indicators, the derived weights were, respectively, 0.182, 0.182, 0.264, 0.104, 0.023, 0.023, 0.016, 0.119, 0.035 and 0.051 with a consistency ratio of 0.037; and for the three impact indicators, were, respectively, 0.751, 0.178 and 0.070 with a consistency ratio of 0.025.

The Vulnerability Index (VI) was built by calculating the weighted sum of the ten variables (16.1):

$$\mathrm{VI}_j = \sum_{i=1}^{n} w_i f_{ij}, i = 1, 2, \cdots, n; j = 1, 2, \cdots, m. \tag{16.1}$$

where, VI_j is the VI for the jth homogeneous unit; w_i is the weight for the ith vulnerability indicator; f_{ij} is the scale for the ith vulnerability indicator in the jth homogeneous unit; m and n are the respective number of homogeneous units and vulnerability indicators. The calculation of Hazard Index (HI) was analogous to (16.1). The Risk Index (RI) was obtained by a simple weighted average of VI and HI. The weights for VI and HI are, respectively, 0.75 and 0.25. Then, the RI was normalized as a percentage of 0–100. Finally, the normalized RI was ranked into five levels: very low (<20%), low (20–40%), moderate (40–60%), high (60–80%) and very high (>80%) (Fig. 16.2).

The results show that the mean of RIs is 45.5%. The values over 50% of RI appear in 17 shoreline segments, amounts to over 35% of the total length of all shorelines of the study area. The shoreline segments with RIs being over 60% are mainly distributed in the four districts of Baoshan, Pudong, Fengxian and Jinshan. Moderate values of RI appear in the southern part of Qidong and Chongming Island, most parts of Changxing Island and Taicang, and the southern part of Nanhui district. The low values appear in Chongming Dongtan, Jiuduansha, and the North Branch.

16.6 Discussion and Conclusions

As Tribbia and Moser (2008) reminds us, there exists a disconnection between science and practice in coastal management. Currently, it is scientists and researchers in the institute who are engaged in the study of coastal erosion, while the managers who need to prepare for the impact of coastal erosion appear not to be involved. In order to assure scientific decision making, managers should adequately

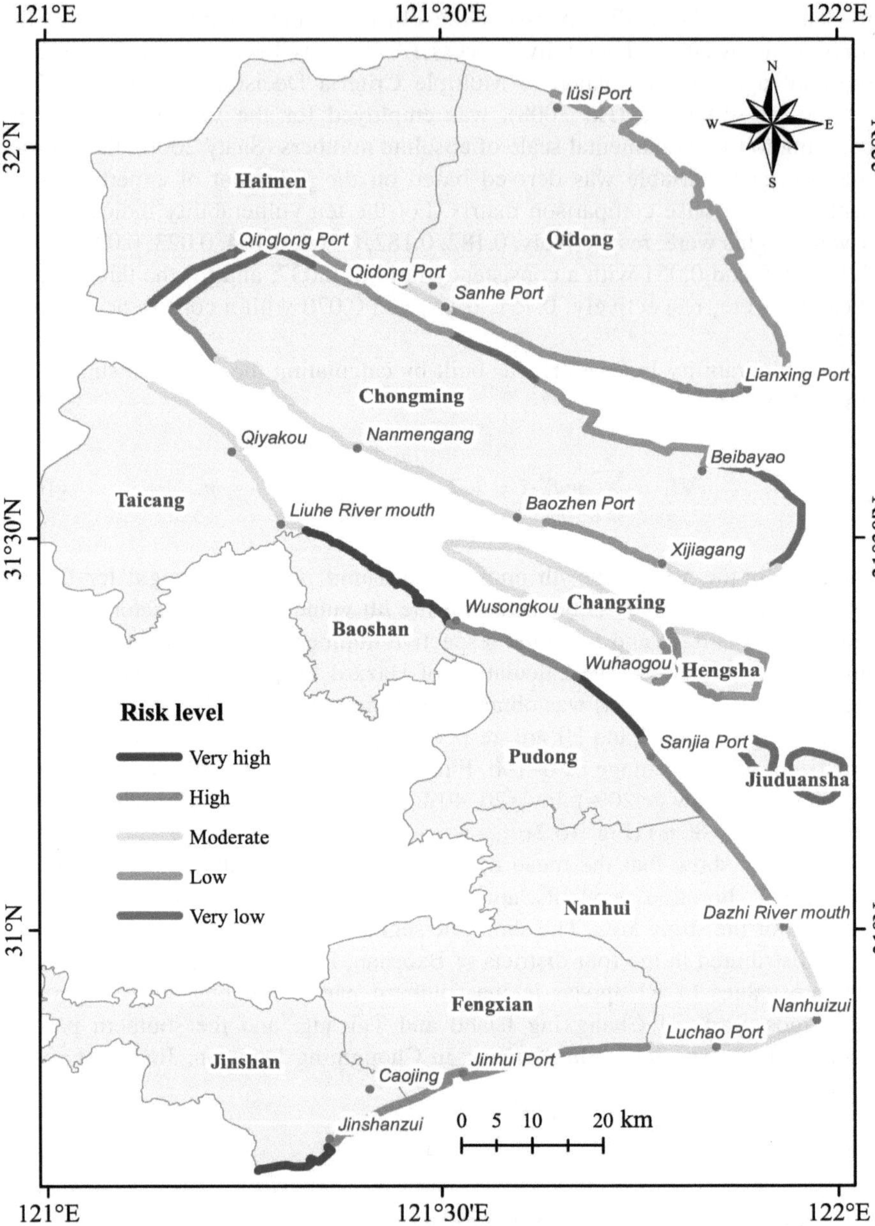

Fig. 16.2 Distribution of risk level for the study area

understand the coastal erosion: the causes, effects and means of modification. There is urgent need for a bridge between scientists and managers. The proposed DSF partially meets these requirements. It has some distinctive characteristics. First of all, with GIS and the linear referencing database system, the framework is scalable and allows the incorporation of a broad range of data and models when such tools are available. The scenario generator and visualization toolkit make communication and synergetic work between humans and computers possible. To implement the framework, however, some challenges still exist.

Data remains one of the biggest challenges to implement the framework in the Yangtze Delta. Although much research has focused on estuarine and coastal environment over the past decades, the data have only been sporadically held and acted upon by individuals and communities. A well-organized data collection and sharing mechanism is strongly suggested. The technique of capturing data is also problematic in highly dynamic environments (Wright 2009). Researchers must rely on the appropriate sampling strategy to guarantee that the inherent characteristic of dynamic elements can be faithfully preserved. A remote sensing observation offers another means of providing useful parameters.

It should be noted that some models, involved in the DSF presented in this paper, inevitably need to address the complexity and uncertainty of coastal systems in the Yangtze Delta. In particular, the results from deterministic models should be carefully considered, because, as claimed by Purvis et al. (2008), this kind of model masks the uncertainty of complex systems and exclude expert knowledge. Dawson et al. (2009) contested that probabilistic methods should be adopted to predict coastal erosion. Indeed for a given model, deterministic model or probabilistic model, the validity of the result depends on many factors (Szlafsztein and Sterr 2007). To minimise the above problems, the scenario generator and visualization toolkit are therefore included in the DSF in this study. Meanwhile, the evolution and involvement of human knowledge are indispensable in the decision making processes of complex systems.

Because of what is considered to be the influence of global warming in terms of such as a rise in sea levels, there is a more urgent need to carry out risk assessment of coastal erosion in the Yangtze Delta area, almost more than any other coastal areas in China. This contention is the stimulus and focus of this study. It has led to the presentation of the DSF for the risk assessment of coastal erosion in Yangtze Delta. A GIS-based risk assessment model for the muddy coasts of the Yangtze Delta has also been developed in this study. The Analytic Hierarchy Process (AHP) method, instrumental in combining computer intelligence and expert knowledge, was used to weight the variables. The assessment results show the validity of the approach. But as described above, the Yangtze Delta region still presents challenges, as regards complete data acquisition and development of high-reliability models, due to the well-known reasons of the developing area. Therefore, there is still some way to go before the proposed DSF can be put into practice.

Local authorities are advised to take heed of the evidence presented in this paper in current work on coastal erosion to better understand the potential of the risk of such erosion and realize that precautions must be taken in advance. Finally, it should be understood that the proposed DSF, especially the vulnerability indicators and the impact indicators, defined for the Yangtze Delta and given in this paper, are equally relevant to a wide range of muddy coastal areas.

Acknowledgements We gratefully acknowledge the support of the Programme Strategic Scientific Alliances between China and the Netherlands (2008DFB90240), the state key laboratory special fund from the Ministry of Science and Technology of China, the School Foundation of Xuzhou Normal University (11XLR04), the Doctoral Fund of Shanghai Ocean University. We also sincerely thank Dr. Ge Jianzhong for his help in simulating significant wave heights. Special thanks are due to anonymous reviewers for their valuable comments and suggestions for the improvement of the manuscript.

References

Alter S (2004) A work system view of DSS in its fourth decade. Decis Support Syst 38(3):319–327

Bird ECF (1985) Coastline changes. Wiley, New York

Cai F, Su X, Liu J et al (2009) Coastal erosion in China under the condition of global climate change and measures for its prevention. Prog Nat Sci 19(4):415–426

Center for International Earth Science Information Network (CIESIN) and Columbia University; United Nations Food and Agriculture Programme (FAO); and Centro Internacional de Agricultura Tropical (CIAT) (2005) Gridded population of the world: future estimates (GPWFE). Palisades: Socioeconomic Data and Applications Center (SEDAC), Columbia University. http://sedac.ciesin.columbia.edu/gpw

Chen X, Zong Y (1998) Coastal erosion along the Changjiang deltaic shoreline, China: history and prospective. Estuar Coast Shelf S 46(5):733–742

Chen JY, Yu ZY, Yun CX (1959) Development of the geomorphology of the Changjiang delta. Acta Ecol Sin (Chinese) 25(3):201–220

Chen JY, Shen HT, Yun CX (1988) Process of dynamics and geomorphology of the Changjiang Estuary (in Chinese). Shanghai Science and Technology Press, Shanghai

Chen XQ, Zong YQ, Zhang EF et al (2001) Human impacts on the Changjiang (Yangtze) river basin, China, with special reference to the impacts on the dry season water discharges into the sea. Geomorphology 41(2–3):111–123

Dawson R, Dickson M, Nicholls R et al (2009) Integrated analysis of risks of coastal flooding and cliff erosion under scenarios of long term change. Clim Chang 95(1):249–288

Del Río L, Gracia FJ (2009) Erosion risk assessment of active coastal cliffs in temperate environments. Geomorphology 112(1–2):82–95

Eurosion (2004) Living with coastal erosion in Europe: Sediment and space for sustainability. guidance document for quick hazard assessment of coastal erosion and associated flooding, Directorate General Environment, European Commission

Feng AP, Xia DX (2003) Grading of coastal erosion disaster situation. Coast Eng (Chinese) 22 (2):60–66

Gao S, Wang YP (2008) Changes in material fluxes from the Changjiang River and their implications on the adjoining continental shelf ecosystem. Cont Shelf Res 28(12):1490–1500

Gong SL (2008) The study on acting factors and systemic control of land subsidence in Shanghai (in Chinese). Ph.D. Thesis, Shanghai: East China Normal University, p 133

Han ZS, Jayakumar R, Liu K et al (2008) Review on transboundary aquifers in People's Republic of China with case study of Heilongjiang-Amur river basin. Environ Geol 54(7):1411–1422

Ho W (2008) Integrated analytic hierarchy process and its applications – a literature review. Eur J Oper Res 186(1):211–228

Hori K, Saito Y, Zhao QH et al (2002) Architecture and evolution of the tide-dominated Changjiang (Yangtze) river delta, China. Sediment Geol 146(3–4):249–264

Jones PD, Tyler AO, Wither AW (2002) Decision-support systems: do they have a future in estuarine management? Estuar Coast Shelf S 55(6):993–1008

Lau M (2005) Integrated coastal zone management in the People's Republic of China–an assessment of structural impacts on decision-making processes. Ocean Coast Manage 48(2):115–159
Li HP, Yang GS (2001) Determination of classification and risk in the coastal zone of Yangtze Delta and North Jiangsu. J Nat Disasters (Chinese) 10(4):20–25
Milliman JD, Huang-ting S, Zuo-sheng Y et al (1985) Transport and deposition of river sediment in the Changjiang estuary and adjacent continental shelf. Cont Shelf Res 4(1–2):37–45
Nicholls RJ, Wong PP, Burkett VR et al (2007) Coastal systems and low-lying areas. In: Parry ML, Canziani OF, Palutikof JP, van der Linden PJ, Hanson CE (eds) Climate change 2007: impacts, adaptation and vulnerability. Contribution of working group II to the fourth assessment report of the intergovernmental panel on climate change. Cambridge University Press, Cambridge, pp 315–356
Power DJ (2008) Decision support systems: a historical overview. In: Burstein F, Holsapple CW (eds) Handbook on decision support systems 1: basic themes. Springer, Berlin, pp 121–140
Purvis MJ, Bates PD, Hayes CM (2008) A probabilistic methodology to estimate future coastal flood risk due to sea level rise. Coast Eng 55(12):1062–1073
Saaty TL (2008) Decision making with the analytic hierarchy process. Int J Serv Sci 1(1):83–98
Saito Y, Yang ZS, Hori K (2001) The Huanghe (Yellow River) and Changjiang (Yangtze River) deltas: a review on their characteristics, evolution and sediment discharge during the Holocene. Geomorphology 41(2–3):219–231
Shen H, Guo C, Zhu H et al (1988) A discussion on the change and origin of turbidity maximum in the Changjiang Estuary. In: Chen J, Shen H, Yu C (eds) Process of dynamics and geomorphology of the Changjiang Estuary (in Chinese). Shanghai Scientific and Technical Publishers, Shanghai, pp 216–228
Sheng JF, Zhu DK (2002) Discussion about coastline erosion and management. Mar Sci Bull (Chinese) 21(4):50–57
Shi YF, Zhu JW, Xie ZR et al (2000) Prediction and prevention of the impacts of sea level rise on the Yangtze River Delta and its adjacent areas. Sci China Ser D 43(4):412–422
Shi C, Hutchinson SM, Yu L et al (2001) Towards a sustainable coast: an integrated coastal zone management framework for Shanghai, People's Republic of China. Ocean Coast Manage 44 (5–6):411–427
Szlafsztein C, Sterr H (2007) A GIS-based vulnerability assessment of coastal natural hazards, state of Para, Brazil. J Coast Conserv 11(1):53–66
Tribbia J, Moser SC (2008) More than information: what coastal managers need to plan for climate change. Environ Sci Policy 11(4):315–328
UN/ISDR (2004). Living with risk: a global review of disaster reduction initiatives, United Nations inter-agency secretariat of the international strategy for disaster reduction, Geneva, Switzerland
Uran O, Janssen R (2003) Why are spatial decision support systems not used? Some experiences from the Netherlands. Comput Environ Urban Syst 27(5):511–26
Van Kouwen F, Dieperink C, Schot P et al (2008) Applicability of decision support systems for integrated coastal zone management. Coast Manage 36(1):19–34
Wang WH, Wu SY, Chen XY (1999) Research on the assessment method of the coastal erosion disaster. J Nat Disasters (Chinese) 8(1):71–77
Westmacott S (2001) Developing decision support systems for integrated coastal management in the tropics: Is the ICM decision-making environment too complex for the development of a useable and useful DSS? J Environ Manage 62(1):55–74
Wiggins S (2004) Coastal decision support systems in the UK. Coast Net Bull 8(3):18
Wright DJ (2009) Spatial data infrastructures for coastal environments. In: Yang X (ed) Remote sensing and geospatial technologies for coastal ecosystem assessment and management. Springer, Berlin, pp 91–112
Xia DX, Wang WH, Wu GQ et al (1993) Coastal erosion in China. Acta Geogr Sin (Chinese) 48 (5):468–476

Yang SL, Ding PX, Chen SL (2001) Changes in progradation rate of the tidal flats at the mouth of the Changjiang (Yangtze) river, China. Geomorphology 38(1–2):167–180

Yang Z, Wang H, Saito Y et al (2006) Dam impacts on the Changjiang (Yangtze) river sediment discharge to the sea: the past 55 years and after the Three Gorges Dam. Water Resour Res 42 (4):W04407

Yang SL, Zhang J, Xu XJ (2007) Influence of the Three Gorges Dam on downstream delivery of sediment and its environmental implications, Yangtze River. Geophys Res Lett 34(10):L10401

Zhang Y, Swift DJP, Yu ZY et al (1998) Modeling of coastal profile evolution on the abandoned delta of the Huanghe River. Mar Geol 145(1–2):133–148

Zhang KQ, Douglas BC, Leatherman SP (2004) Global warming and Coastal Erosion. Clim Chang 64(1):41–58

Zhou D, Liang YB, Zeng CK (1994) Oceanology of China seas, vol II. Kluwer Academic Publishers, Dordrecht/Boston

Zuo X, Aiping F, Ping Y et al (2009) Coastal erosion induced by human activities: a Northwest bohai sea case study. J Coast Res 25(3):723–733

Chapter 17
Estimating Ice Thickness in South Georgia from SRTM Elevation Data

A. Paul R. Cooper, James W. Tate, and Alison J. Cook

Abstract South Georgia is a glaciated island in the South Atlantic, which provides a primary nesting site for the albatrosses and petrels of the Southern Ocean. 60% of the island is covered by glaciers and ice fields, and the majority of the coastal glaciers are observed to be retreating. A small number of these glaciers are advancing, and others are retreating at anomalously fast rates. As the status of these glaciers is important for environmental management of South Georgia, potentially controlling the spread of invasive species into currently pristine regions, it is necessary to understand the pattern of glacier change in South Georgia. However, detailed study of the glaciology of South Georgia is hampered by lack of measurements of the thickness of the ice. Because of the logistic difficulties of operating on South Georgia, there are no conventional ice thickness measurements from drilling, radar or seismic techniques, and it is unlikely that these will be available in the near future.

This paper addresses this lack of basic information by using surface slope data to estimate the ice thickness of glaciers and ice fields in South Georgia. The surface slope data are derived using surface elevations from the Shuttle Radar Topographic Mission, which provides elevation measurements with a high relative accuracy. The estimate of ice thickness critically depends on assumptions about the conditions in the ice column and at the base of the ice mass, and areas where the estimate is clearly in error provide an insight into changed ice flow conditions or the environment at the ice/rock interface. These anomalous regions are then compared with

A.P.R. Cooper (✉) • A.J. Cook
British Antarctic Survey, High Cross, Madingley Road, Cambridge CB3 0ET, UK
e-mail: paul.cooper@bas.ac.uk; acook@bas.ac.uk

J.W. Tate
Department of Civil, Environmental and Geomatic Engineering, University College London, London, UK
e-mail: james.tate@ucl.ac.uk

A. Yeh et al. (eds.), *Advances in Spatial Data Handling and GIS*, Lecture Notes in Geoinformation and Cartography,
DOI 10.1007/978-3-642-25926-5_17,

glacier change data, providing insights into the reasons for the unusually rapid retreat or advance of certain glaciers.

The paper describes the methodology used to compute ice thickness values, with an estimate of accuracy and variability of the thickness measures under different assumptions. The paper then identifies regions with anomalous thickness measurements, and seeks to ascertain why the thickness measurement is unreliable in certain regions. Finally these anomalous areas are compared with coastal change data to suggest why certain glaciers are retreating or advancing more rapidly than the norm for South Georgia, and to make predictions concerning future glacier change.

17.1 Introduction

The island of South Georgia is located in the South Atlantic centred on 54° S, 37° W (Fig. 17.1). It is isolated and heavily glaciated with 60% of the island being covered by glaciers or ice sheets. Its location and isolation make it the primary breeding ground for many species of marine birds, such as albatrosses, petrels and penguins.

Recent studies (Cook et al. 2010) have shown that the majority of coastal glaciers on South Georgia are in retreat. A small minority are, however, advancing, and a further small number are retreating far more rapidly than the average. As Cook et al. (2010) show, these changes in glacier extent may have significant consequences on the breeding success of these iconic birds. Introduced terrestrial predators, in particular rats, are currently blocked from major breeding grounds by glacier barriers, and continuing retreat of glaciers threatens these breeding grounds.

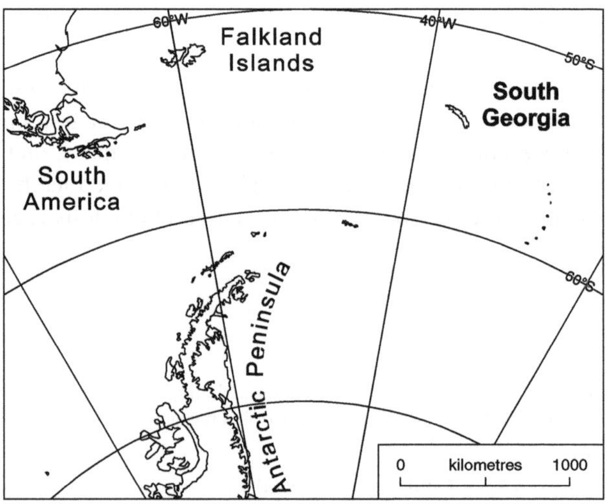

Fig. 17.1 Location map for South Georgia

Comparison of areas occupied by rats and those not shows that most species of bird nesting in South Georgia cannot breed successfully in areas inhabited by rats.

In order to understand glacier dynamics, a crucial parameter is ice thickness. This can be measured in a variety of ways using standard techniques such as seismic sounding, ice-penetrating radar or even by drilling. However, all these techniques require substantial logistic support, which is not available in South Georgia. South Georgia is only accessible by ship; there is no landing ground for fixed-wing aircraft. While an over-snow expedition could potentially carry out a survey of ice thickness, it would be limited in its areal coverage compared with airborne survey. Aircraft equipped for ice-penetrating radar surveys do not have sufficient range to perform a survey over South Georgia after flying from the nearest airfield at Port Stanley in the Falkland Islands.

Fortunately, South Georgia was covered by the Shuttle Radar Topography Mission (SRTM) during February 2000, being just north of the southern limit of 56° S. The product used was DTED 1, 3 arc-second product (~90 m post spacing). SRTM elevations have an absolute accuracy of 8 m (90% probability) for islands, and a relative accuracy of 6.2 m (Farr et al. 2007). Farr et al. also state that accuracies are worst over steep slopes, and better for flat areas, such as glaciers and ice fields, though the improvement of accuracy over flatter areas is not quantified. These relatively accurate elevation data permit an estimate of ice thickness to be made from the surface slope of the glaciers (Paterson 1981, p. 86).

In places, the ice thickness estimates are clearly substantially in error, giving unrealistically high estimates. This can be linked to changes at the base of the glacier, providing an insight into conditions that are relevant to the pattern of retreat of coastal glaciers.

17.2 Method

17.2.1 Theory

The surface slope of a glacier in a steady state is related to the ice thickness by the following relationship:

$$\tau = \rho \cdot g \cdot h \cdot \sin \alpha \qquad (17.1)$$

where τ is the basal shear stress, ρ is the density of ice, g is the acceleration due to gravity, h is the ice thickness and α is the surface slope (Paterson 1981, p. 86).

This equation can be re-arranged to provide a relationship between ice thickness and surface slope, assuming a constant basal shear stress:

Table 17.1 Table of corrections for a glacier with a parabolic cross-section (Paterson 1981, p. 103)

W	F (Parabola)
1	0.445
2	0.646
3	0.746
4	0.806
	1.000

$$h = \frac{\tau}{\rho \cdot g \cdot \sin \alpha} \tag{17.2}$$

So, assuming that the retarding forces at the base of the glacier (τ) are unvarying, it is possible to estimate ice thickness using surface slope values alone, as all other terms in the equation are constant. The value of τ varies within the range 50–150 kPa depending on a variety of factors including the temperature of the ice and the nature of the substrate; a reasonable assumption for its value in the absence of other information is therefore 100 kPa.

Glaciers in South Georgia are constrained by valley walls, so additional corrections are required to account for this.

$$h = \frac{\tau}{\rho \cdot g \cdot \sin \alpha \cdot F} \tag{17.3}$$

Where F is a correction factor that depends on W, the ratio of the distance to the valley wall and the ice thickness on the centre-line of the glacier. GIS techniques detailed below allowed the distance to the valley wall to be computed accurately, and F was obtained from Table 17.1.

Given that much of South Georgia is covered by perennial snow or ice, determining the location of glacier margins is not trivial. A variety of techniques were tested, but the most reliable was clipping the slope data at a value of 17°. Methods based on image analysis using a composite Landsat ETM + image failed due to snow cover on glaciers and heavy shadowing, but were used to eliminate areas of low slope that are not snow-covered (e.g. deglaciated areas in front of retreating glacier snouts). The second derivative of the surface (i.e. rate of change of slope) in many areas provided a good delineation of the edge of a glacier, but failed in areas where the glacier merged into snow-fields and at ice-falls. Having determined the glacier margins, the next step was to compute the glacier centrelines. This was done by computing the Euclidean distance from the glacier margins, the centre-line is then the trace of the maximum distances from the glacier margins. The distance to the glacier wall is then available at every point along the glacier centre-line, and the mean ice thickness can be computed by averaging over a small region along the centre-line. These parameters are used to compute the correction factor in (17.3) above.

17.2.2 *Assumptions*

In order to estimate ice thickness using the technique described above, certain assumptions are made. These are:

1. That the glacier is flowing by plastic deformation.
2. That ice motion is primarily horizontal.
3. That the limiting basal stress is constant.
4. That the value of the limiting basal stress is 100 kPa.
5. That the cross-section of the glacier is approximated by a parabola.

A further issue is that there is a likely (but not precisely known) difference in the surface measured by the SRTM radar between ablation and accumulation areas of the glacier, caused by differences in the density of firn and glacier ice. However, this will not affect the overall results of this analysis, as the surface density of the ice changes slowly with horizontal location.

These assumptions all break down under certain circumstances, which will be discussed further below. However, assumption 4 is unavoidable; there is no way of determining the actual limiting shear stress. 100 kPA is a reasonable value; measured values (computed for glaciers where the ice thickness is known) vary over a range from 50–150 kPa (Paterson 1981, p. 86); values computed for the Antarctic ice sheets show lower values in certain areas where it is likely that the base of the glacier is lubricated (Drewry 1983, Sheet 5). Assumption 5 is made for convenience; the correction parameter for valley glaciers (F in (17.3)) has only been computed for a limited range of cross-sectional shapes, and a parabola is the most realistic. However, computations of the correction factor for other cross-sectional profiles only vary by about 10% from each other; errors from breakdown of this assumption will be small compared with other sources of error.

17.3 Results

The results of the estimation of ice thickness are shown in Fig. 17.2 and summarized in Table 17.2. As breakdown of the assumptions detailed in Sect. 17.2.2 results in overestimates of ice thickness in almost all cases, high ice thicknesses are more likely to be incorrect than lower ice thickness estimates. Figure 17.3 shows an area where surface slopes are very low in the accumulation area of several glaciers; as can be seen from the 100 m contours of ice thickness, the apparent thickness gradient increases rapidly as the ice thickness increases, being so great when estimated ice thickness exceeds 1,000 m that contours above 1,000 m have been omitted. These high and rapidly changing slopes are not found in deglaciated terrain, and are unlikely to be real. Therefore, the likelihood of the estimate being correct decreases with increasing estimated ice thickness. Similarly, high ice thicknesses in the lower parts of certain glaciers, as shown in Figs. 17.7,

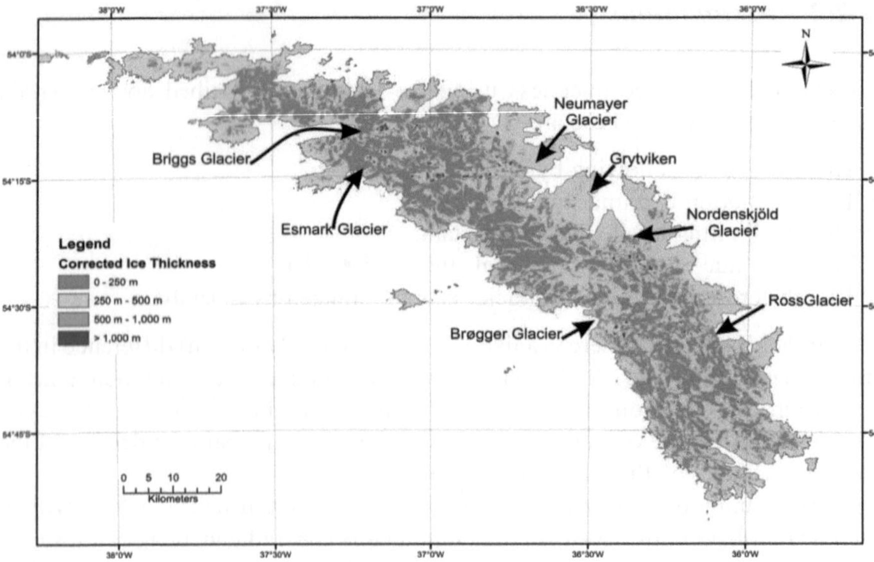

Fig. 17.2 Estimated ice thickness, with locations mentioned in the text

Table 17.2 Area of ice thickness classes

Thickness band	Number of pixels	Area (km^2)	% of ice area
0–250 m	104,861	849	74.57%
250–500 m	26,321	213	18.72%
500–1,000 m	7,365	60	5.24%
>1,000 m	2,071	17	1.47%

17.8, 17.9 and 17.10, are also correlated with high thickness gradients, again suggesting that these high estimates are caused by breakdown of the assumption that the glacier is closely coupled to its bed and is moving by plastic deformation.

In Fig. 17.2, ice thicknesses have been divided into four colour bands corresponding to the likelihood of the results being correct. Inspection of ice thickness gradients suggests that ice thickness estimates between 0 and 250 m are likely to be an accurate reflection of the actual ice thickness (green on Fig. 17.2), estimates between 250 m and 500 m (yellow on Fig. 17.2) are probably subject to errors in some locations; estimates between 500 m and 1,000 m (orange on Fig. 17.2) are very likely to be incorrect, and estimates exceeding 1,000 m (red on Fig. 17.2) are highly unlikely to be correct.

Regions where the estimate is likely to be unreliable are concentrated in two general regions: inland zones crossing ice-sheds, and the coastal ends of some large glaciers. In both cases, the unreliability is caused by break-down of the basic assumptions of the technique.

Areas in the vicinity of ice-sheds are unreliable because ice motion in these regions has a strong vertical component, defeating the simple analysis of Sect. 17.2.1.

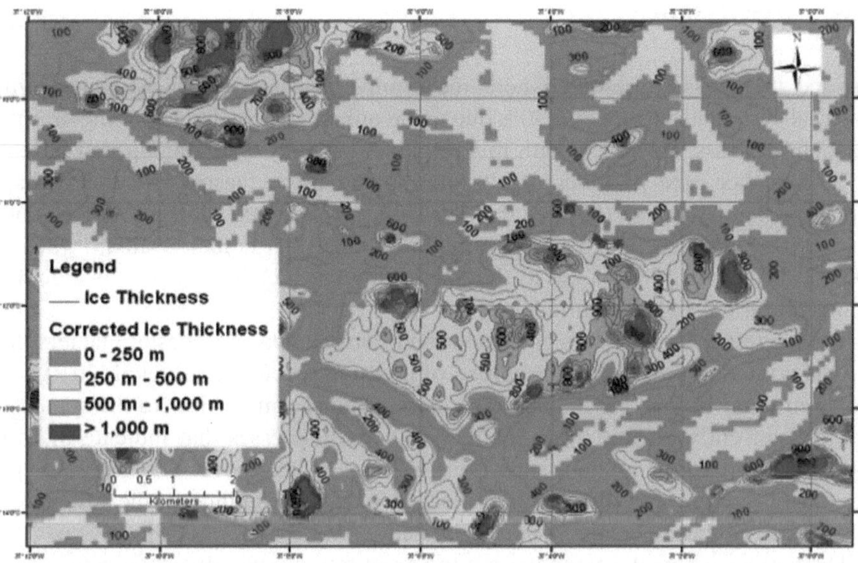

Fig. 17.3 Ice thicknesses illustrating errors

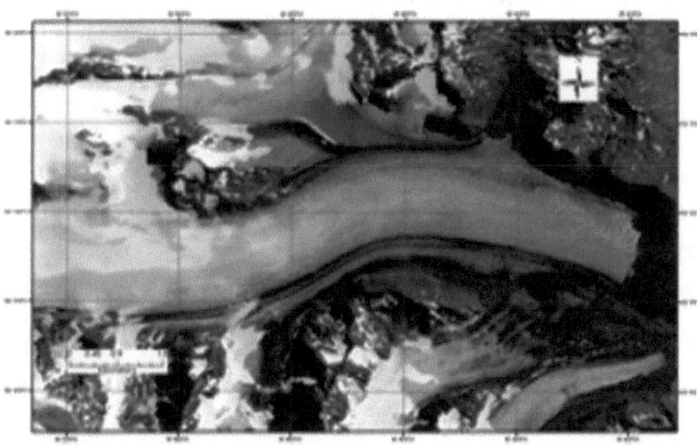

Fig. 17.4 Landsat ETM image of Neumayer Glacier showing flowlines

Areas in the coastal regions of glaciers are almost certainly lubricated at the bed, and the limiting basal shear stress is substantially reduced from the nominal value of 100 kPa. Furthermore, the assumption of plastic flow may also be untrue in this area; the ice may be moving as a block sliding over the base. This is shown by examination of satellite images, which show evidence of rapid streaming flow evidenced by the presence of flow-lines (Fig. 17.4). The stream-lines can be better seen in this link (Google Maps), which provides access to high resolution copyright images.

17.4 Discussion

The ice thicknesses presented here must only be regarded as estimates, with potential systematic errors of ±50% due to the unknown actual value of the limiting basal shear stress. However, with this proviso, it is very encouraging that at least 75% of the ice area provides results that appear to be reasonable, and perhaps another 10% of the area (50% of the 250–500 m) band may well also be reliable. Comparison with ice thicknesses from similar sized but colder Svalbard glaciers (Dowdeswell et al. 1984) suggests that the ice thickness estimates of a few hundred metres are of the right magnitude.

The ice thicknesses of Briggs Glacier and Esmark Glacier, coastal glaciers that may be critical to preventing the spread of invasive species (Cook et al. 2010), are shown in Figs. 17.5 and 17.6. These glaciers, which form barriers to the spread of rats into currently rat-free areas, both show ice-thicknesses that indicate that the glacier flow conforms to the assumptions of Sect. 17.2.2 above. This suggests that previously rapid retreat of these two glaciers may not continue in the future, and that the bed of these glaciers is not lubricated.

The regions where the estimates are too great fall into two distinct classes. First, regions at or near the ice-sheds, where the horizontal motion of the glacier is small, and the ice velocity vectors have a strong downwards component. In these regions, the ice motion does not correspond to the simple model of Sect. 17.2.1, and so the ice thickness estimates are too great. However, in these regions (where ice thickness

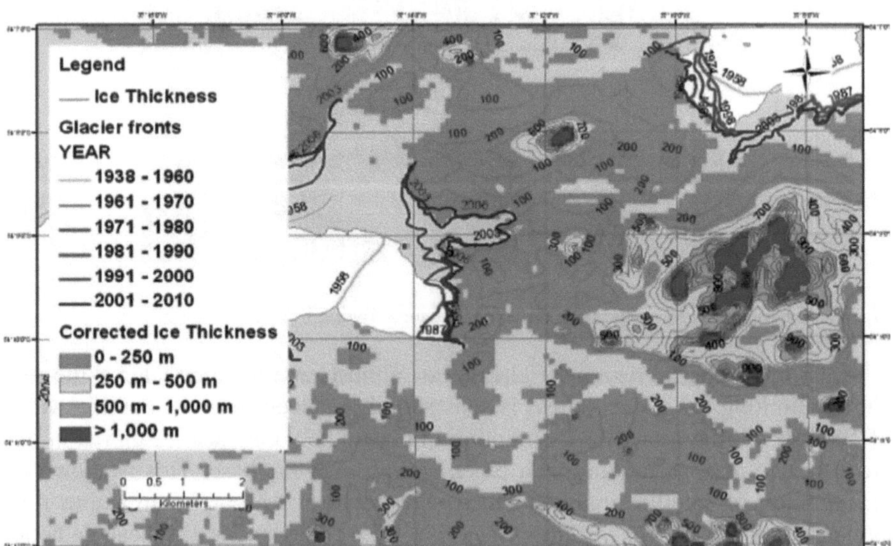

Fig. 17.5 Briggs Glacier showing ice thickness estimates and glacier retreat

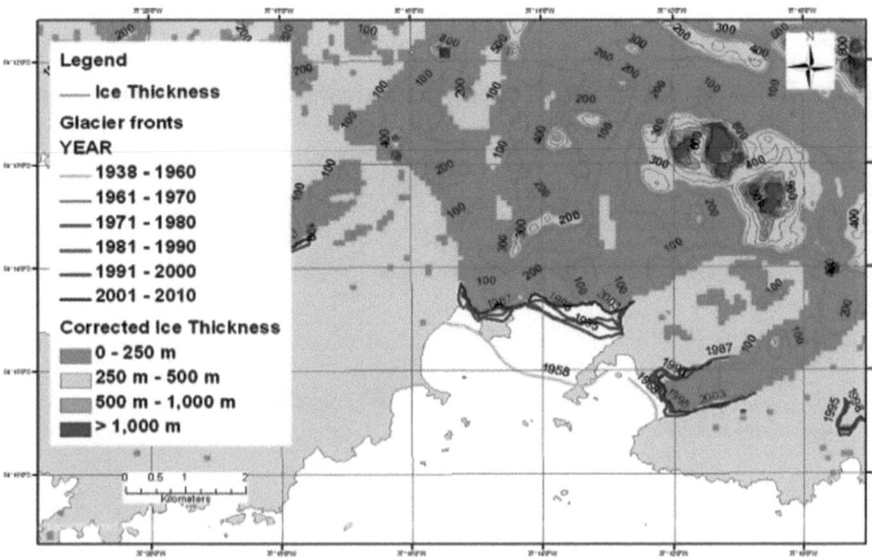

Fig. 17.6 Esmark Glacier showing ice thickness estimates and glacier retreat

is expected to be at its greatest), the estimates in the range 250–500 m are more likely to be correct.

More interesting are the regions in the coastal extremities of certain large glaciers. As noted above, it is likely that in these regions the increased thickness estimates are due to decoupling of the glacier from the bed, reducing the limiting basal shear stress, and so causing an over-estimate of the ice thickness. This decoupling is likely to be caused by the presence of water at the bed of the glacier. The darker ice shown in Fig. 17.4 indicates that the apparently high thickness region on Neumayer Glacier corresponds to the ablation zone of the glacier, as well as showing other evidence of bed decoupling in the form of stream-lines.

Of the major glaciers of South Georgia, two on the north coast (Neumayer Glacier (Fig. 17.7) and Ross Glacier (Fig. 17.8)) showing this bed decoupling also show rapid retreat. Neumayer Glacier has retreated by over 4 km and Ross Glacier by 3 km. Nordenskjöld Glacier (Fig. 17.9), in a similar setting, has retreated, but only by 850 m, a much smaller retreat. A fourth, Brøgger Glacier (Fig. 17.10), on the south coast, shows no retreat at present.

It is noteworthy that the three glaciers showing retreat are all on the north coast of South Georgia, and because of prevailing westerly wind direction, the orographic effect of the mountains forming the central spine of the island will cause a lower accumulation rate than that to the south of the central mountains. Any reduction in the accumulation rate or increase in the elevation of the equilibrium line due to increasing temperatures will cause rapid retreat, especially where the glacier velocity is high. The temperatures at Grytviken, on the north coast of South Georgia, show

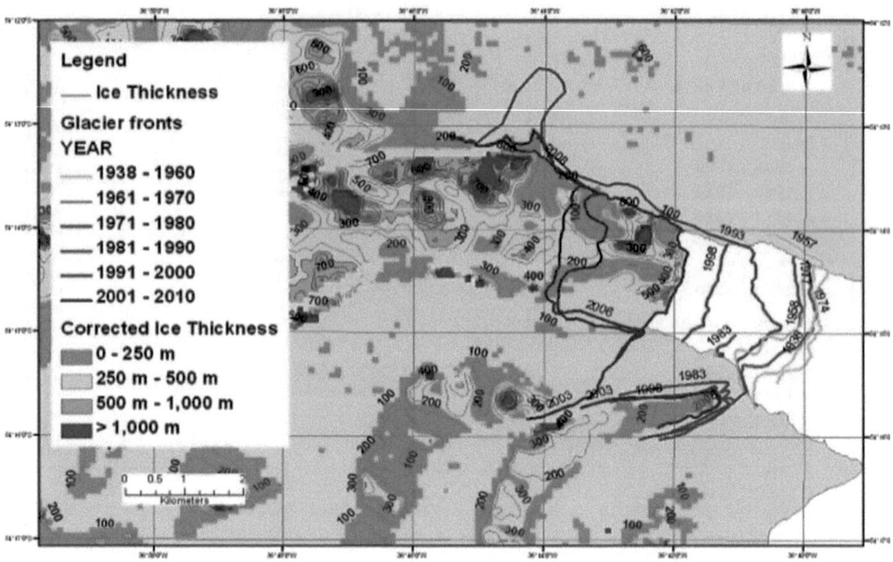

Fig. 17.7 Neumayer Glacier showing ice thickness estimates and glacier retreat

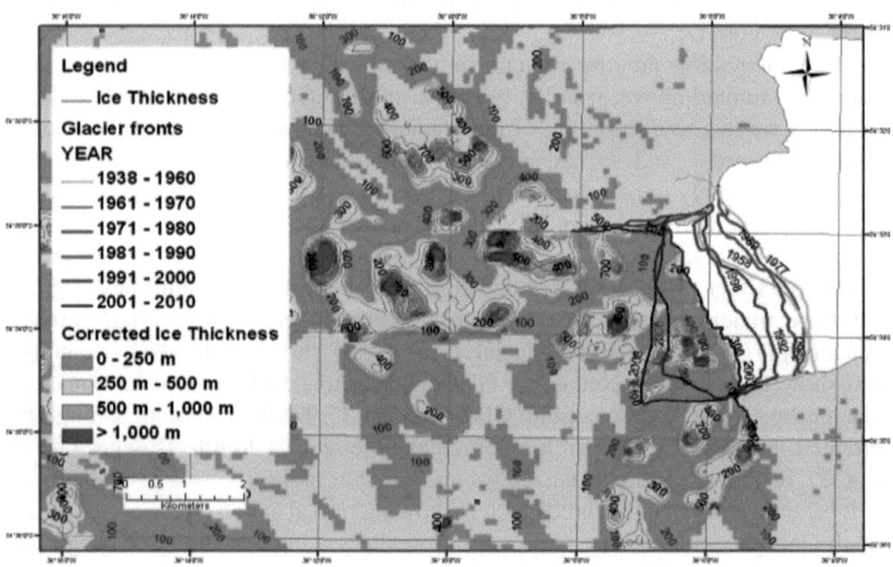

Fig. 17.8 Ross Glacier showing ice thickness and glacier retreat

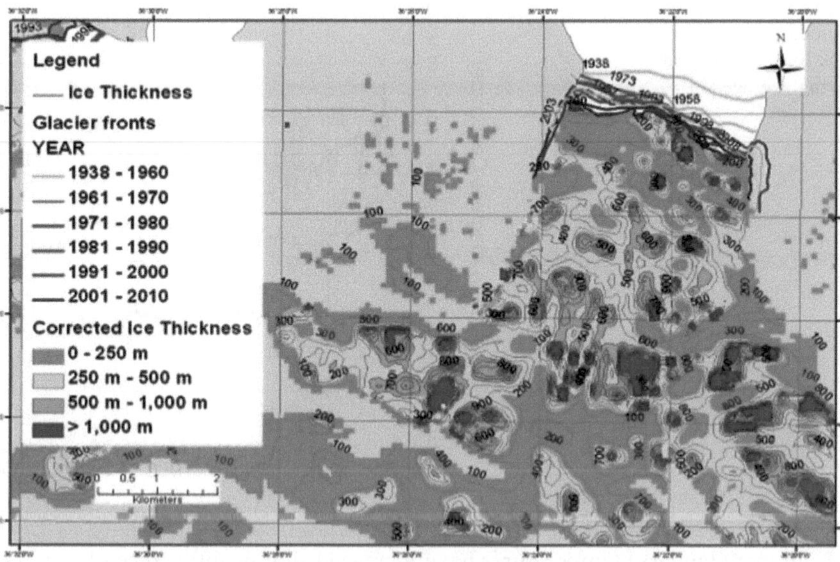

Fig. 17.9 Nordenskjöld Glacier showing ice thickness and glacier retreat

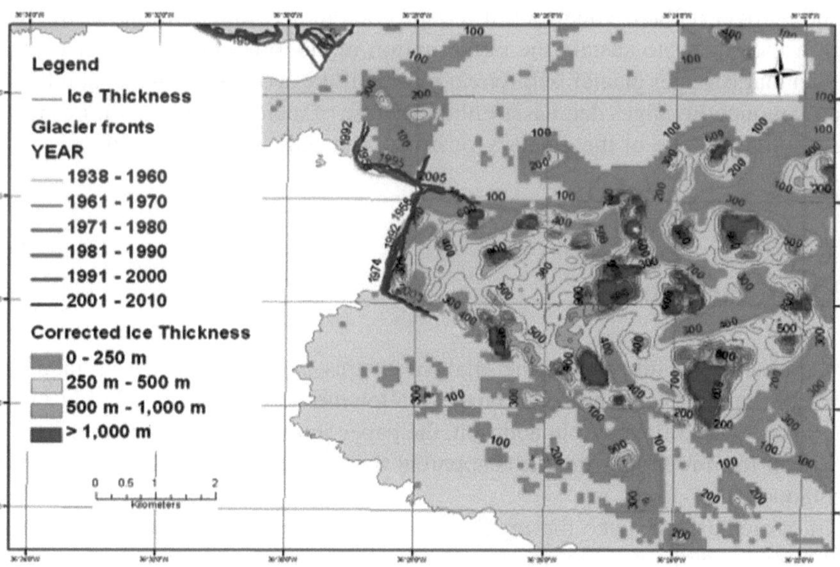

Fig. 17.10 Brøgger Glacier showing ice thickness and glacier retreat

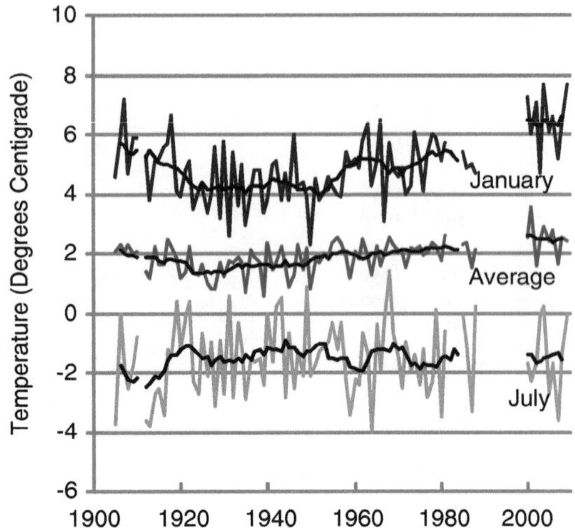

Fig. 17.11 Temperatures recorded at Grytviken. The black lines are 10 year moving averages

that summer (January) temperatures have increased significantly since observations began in 1905, and are nearly 2° higher than the minimum temperatures observed in the 1930s (Fig. 17.11). Winter temperatures show little, if any, change.

The anomalous ice thickness values for Brøgger Glacier suggest that the glacier is sliding over its bed. Therefore, it requires a high accumulation rate in the upper part of the glacier to sustain the rapid motion in the lower part of the glacier. We predict that Brøgger glacier will retreat rapidly if the accumulation rate south of the central mountain ranges decreases either because of increasing temperatures raising the equilibrium line of the glaciers, or because of reduced precipitation.

17.5 Conclusions

In the absence of other data, the analysis of this paper shows that surface slope derived from remotely-sensed surface elevation measurements can provide a useful estimator for ice thickness. In addition, the paper shows that useful conclusions can be drawn from regions where it is apparent that the underlying assumptions of the technique break down.

In particular, we make the prediction that Nordenskjöld and Brøgger Glaciers will retreat rapidly at some point in the future, as increasing global temperatures reduce the accumulation available to drive these glaciers.

Acknowledgements The authors wish to express their appreciation for input from their colleagues, especially Mr A H Fleming and Mr D J Herbert of BAS. Mr Tate wishes to express his thanks to Professor I Dowman, his supervisor at University College, London.

References

Cook AJ, Poncet S, Cooper APR, Herbert DJ, Christie D (2010) Glacier retreat on South Georgia and implications for the spread of rats. Antarct Sci 22(3):255–263

Dowdeswell JA, Drewry DJ, LiestøL O, Orheim O (1984) Radio echo-sounding of Spitsbergen glaciers: problems in the interpretation of layer and bottom returns. J Glaciol 30(104):16–21

Drewry DJ (1983) Antarctica: glaciological and geophysical folio. Scott Polar Research Institute, Cambridge

Farr TG et al (2007) The shuttle radar topography mission. Rev Geophys 45, RG2004. doi: 10.1029/2005RG000183

Paterson WSB (1981) The physics of glaciers. Pergamon, Oxford, p 380

